Urban Food Systems Governance and Poverty in African Cities

As Africa urbanises and the focus of poverty shifts to urban centres, there is an imperative to address poverty in African cities. This is particularly the case in smaller cities, which are often the most rapidly urbanising, but the least able to cope with this growth. This book argues that an examination of the food system and food security provides a valuable lens to interrogate urban poverty. Chapters examine the linkages between poverty, urban food systems and local governance with a focus on case studies from three smaller or secondary cities in Africa: Kisumu (Kenya), Kitwe (Zambia) and Epworth (Zimbabwe).

The book makes a wider contribution to debates on urban studies and urban governance in Africa through analysis of the causes and consequences of the paucity of urban-scale data for decision makers, and by presenting potential methodological innovations to address this paucity. As the global development agenda is increasingly focusing on urban issues, most notably the urban goal of the new Sustainable Development Goals and the New Urban Agenda, the work is timely.

Jane Battersby is a senior researcher at the African Centre for Cities, University of Cape Town, South Africa, and is the Research Coordinator of the ESRC/DFID-funded Consuming Urban Poverty project and PI of the IDRC-funded Nourishing Spaces Project. She is the Laureate of the Premio Daniel Carasso 2017. An urban geographer by training, her work focusses on urban food security, food systems and governance.

Vanessa Watson is a Professor of City Planning and Fellow of the University of Cape Town, South Africa. She holds degrees, including a PhD, from South African Universities and the Architectural Association of London, UK, and is on the executive committee of the African Centre for Cities. She is the PI of the ESRC/DFID-funded Consuming Urban Poverty project.

Routledge Studies in Food, Society and the Environment

Peasants Negotiating a Global Policy Space
La Vía Campesina in the Committee on World Food Security
Ingeborg Gaarde

Public Policies for Food Sovereignty
Social movements and the state
Edited by Annette Desmarais, Priscilla Claeys and Amy Trauger

Sustainable Food Futures
Multidisciplinary solutions
Edited by Jessica Duncan and Megan Bailey

Food Riots, Food Rights and the Politics of Provisions
Edited by Naomi Hossain and Patta Scott-Villiers

Food Sovereignty, Agroecology and Biocultural Diversity
Constructing and contesting knowledge
Edited by Michel Pimbert

Food and Nutrition Security in Southern African Cities
Edited by Bruce Frayne, Jonathan Crush and Cameron McCordic

The Real Cost of Cheap Food
(Second Edition)
Michael Carolan

Food Bank Nations
Poverty, Corporate Charity and the Right to Food
Graham Riches

Urban Food Systems Governance and Poverty in African Cities
Edited by Jane Battersby and Vanessa Watson

For further details please visit the series page on the Routledge website: www.routledge.com/books/series/RSFSE/

Urban Food Systems Governance and Poverty in African Cities

Edited by Jane Battersby
and Vanessa Watson

Routledge
Taylor & Francis Group
LONDON AND NEW YORK

earthscan
from Routledge

First published 2019
by Routledge
2 Park Square, Milton Park, Abingdon, Oxon OX14 4RN

and by Routledge
52 Vanderbilt Avenue, New York, NY 10017

First issued in paperback 2020

Routledge is an imprint of the Taylor & Francis Group, an informa business

British Library Cataloguing-in-Publication Data
A catalogue record for this book is available from the British Library

Library of Congress Cataloging-in-Publication Data
Names: Battersby, Jane, 1975– editor, author. | Watson, Vanessa, editor.
Title: Urban food systems governance and poverty in African cities / edited by Jane Battersby and Vanessa Watson.
Other titles: Routledge studies in food, society and environment.
Description: Routledge : New York, 2018. | Series: Routledge studies in food, society and the environment | Includes bibliographical references and index.
Identifiers: LCCN 2018010500 | ISBN 9781138726758 (hardback) | ISBN 9781315191195 (ebook)
Subjects: LCSH: Food security—Africa, Sub-Saharan. | Food supply—Government policy—Africa, Sub-Saharan. | Poverty—Africa, Sub-Saharan. | Urbanization—Africa, Sub-Saharan.
Classification: LCC HD9017.A3572 U73 2018 | DDC 338.1967—dc23
LC record available at https://lccn.loc.gov/2018010500

ISBN 13: 978-0-367-58756-7 (pbk)
ISBN 13: 978-1-138-72675-8 (hbk)

Typeset in Bembo
by Apex CoVantage, LLC

Contents

Figures

Tables

Boxes

Foreword

This book presents the findings of an international collaborative research project that aimed to improve our understanding of the connections between urban poverty, food systems, household food security and governance, by focusing on three secondary cities in Anglophone sub-Saharan Africa.

On the whole, colonial governments sought to prevent people from moving into urban areas, not least because of the potentially politically destabilising effects of population concentrations. Where urban workers were needed, the colonial powers developed urban policies and governance systems that were strongly influenced by European ideas about well-planned towns and cities and public health concerns but disenfranchised urban residents. In these contexts, colonial governments did indeed pay attention to food, through measures that sought to ensure a supply of cheap food for wage workers (a continuing priority for post-independence governments). The reality of increasing rural–urban migration and governance arrangements that were incapable of keeping pace with the demands of low-income urban growth were soon clear, but governments were reluctant to admit the shortcomings of their own policies and urban management systems.

Many aspects of urban growth that were the subject of research and policy development in richer countries, including those from which African countries won independence between the 1960s and the 1980s, were neglected, so the evidential basis for developing better informed and more appropriate policies was lacking, while policies driven mainly by the interests of national politicians and, in these aid-dependent countries, international financial institutions failed to fully acknowledge urban problems and to tackle them coherently. My own research in Lusaka in the 1970s revealed that urban agriculture was a widespread practice, while the eventual publication of that evidence sought to make visible an activity that seemed to be invisible to researchers, drawing attention to its gendered characteristics in a context where rural women had primary responsibility for household food production (Rakodi 1988). By that time, more attention was being given to the historical evolution of food supply and policies (Guyer 1987), and new research on urban agriculture was being undertaken.

By the 1980s, the devastating results of economic liberalisation policies for countries attempting to diversify their economies away from agriculture and mining through protecting nascent manufacturing sectors were becoming clear, but attention focused on the implications of changes in agricultural support and pricing policies for rural poverty, while national policy makers and international donor agencies alike were reluctant to acknowledge the adverse effects of structural adjustment policies for urban household livelihoods and urban management capacity. The focus of research shifted to developing improved analytical frameworks for understanding urban household livelihoods and documenting urban poverty. From the outset, it was recognised that household livelihoods cannot be understood in isolation from the wider economic and political linkages that provide or constrain household choices related to income earning, consumption, land and housing tenure, access to services, and political participation. For some years, too little attention was given to these wider links, including the environmental and governance systems that influence water supply and waste management, energy supply and, indeed, food. Gradually, more attention has been given to them and it is in the latter area that this book makes its path-breaking contribution.

The Consuming Urban Poverty research project builds on longstanding but patchy research on urban planning and management; politics at the local, city and national levels; household livelihoods; 'informal sector' activities; and urban land supply and management, as well as more recent related research on food systems and food security, using innovative methods such as commodity value chain analysis. Its in-depth studies assembled new data on the case study cities of Kisumu in Kenya, Kitwe in Zambia and Epworth in Zimbabwe, situated within a review of the wider sub-Saharan African policy and data scenes. It reveals the lived reality of food poverty for many urban households, including a lack of dietary diversity and frequent food insecurity. It stresses the continuing rural bias of national and international food security policy, one of the reasons why interventions are often inappropriate and damaging. It unpicks some of the complex interconnections between poverty, food systems and governance, and shows that urban food systems have intra-urban, city-regional, national, and international dimensions. Revealing the many links between formal and informal food supply systems, it strengthens calls for improved theoretical understanding of urban economies. And it documents how limited and inaccurate understanding of the complex interconnections between poverty, food systems and governance, professional preconceptions and political dynamics leads to interventions that are often inappropriate and damaging.

I was delighted to be invited to chair the Advisory Board for this project. The Board was a small, knowledgeable and committed group of people who brought international academic and donor perspectives, knowledge of and connections with the local and national contexts of the three case study cities, and expertise in research communication to advise the teams at the African Centre for Cities and in the three case study cities. I hope that our inputs contributed positively to their work.

I look forward to this team and others taking forward the research, policy and practice work identified. This research shows that future work needs to overtly acknowledge and analyse the realities of political power relationships, examine the drivers and characteristics of changes in urban food wholesaling and retailing systems, and ensure that analysis and policy is gendered throughout. This study has contributed to the evolution of theoretical and analytical frameworks for understanding city economies, urban land tenure and management, and the interactions between local, city and national politics, but further refinements to ways of studying and understanding city economies, urban land tenure, management and planning, and the interactions between local, city and national politics are required. In addition, there is a need to evaluate the outcomes of policy and planning interventions and to assess their transferability to urban and country contexts with different colonial histories, economic bases, political configurations, governance capacities and vulnerability to climatic and environmental strains and stresses.

Carole Rakodi, Emeritus Professor, International Development Department, School of Government and Society, University of Birmingham

References

Guyer, J. (ed.) (1987) *Feeding African Cities: Studies in Regional and Social History*. Manchester: Manchester University Press for the International African Institute, London.

Rakodi, C. (1988) Urban agriculture: research questions and Zambian evidence. *The Journal of Modern African Studies*, 26(3), pp. 495–515.

Contributors

Frankline Otiende Awuor is a lecturer in the School of Spatial Planning and Natural Resources Management at Jaramogi Oginga Odinga University of Science and Technology (JOOUST). Dr. Otiende is a range ecologist, having graduated from the University of Nairobi with a BSc and MSc in Range Management and a PhD in Planning from JOOUST. Currently, he is a post-doctoral fellow in food security at the Kisumu Local Interaction Platform (KLIP). His research interests are in solid waste management, wildlife management, and ecotourism.

Harro von Blottnitz is a professor in the Faculty of Engineering and the Built Environment at the University of Cape Town (UCT), where he is based in the highly dynamic and successful Chemical Engineering Department. Professor von Blottnitz holds a BSc in Chemical Engineering from UCT, a BSc Honours in Operations Research from UNISA, an MSc in Engineering from UCT, and a Doctorate in Engineering from the RWTH Aachen in Germany. He defines his research and teaching interests by the multiple challenges of sustainable development in developing country settings and has published widely in the fields of Life Cycle Assessment, Renewable Energy (biogas, biodiesel and bio-ethanol), and Waste Management.

Jacqueline Borel-Saladin completed her undergraduate, Honours, and Masters degrees in Environmental and Geographical Science and her PhD in Sociology, with a particular focus on changing employment patterns in South Africa, at the University of Cape Town (UCT). She then moved to the Human Sciences Research Council (HSRC), where she worked on topics ranging from resilient cities and the green economy to the state of secondary cities in South Africa and national urban planning policy in Africa. She is currently a post-doctoral research fellow on the Consuming Urban Poverty project at the African Centre for Cities (UCT), focusing on urbanisation and poverty in the project's case study cities.

Easther Chigumira is a senior lecturer in the Department of Geography and Environmental Sciences at the University of Zimbabwe. She holds a

BA(Hons) and Masters of Social Science degree in Geography from Rhodes University, and a PhD in Geography (specialization in Political Ecology) from the University of Oregon (USA). She is a political ecologist with research and capacity building interests in land and agrarian studies, urban food studies, climate change resilience, and the green economy. She is a country partner coordinator for the ESRC/DFID-funded research project Governing Food Systems to Alleviate Poverty in Secondary Cities in Africa.

Jonathan Crush is the CIGI Chair in Global Migration and Development at the Balsillie School of International Affairs. He co-founded and directs the Hungry Cities Partnership and African Food Security Urban Network (AFSUN) and is a co-investigator on the Consuming Urban Poverty project at the African Centre for Cities. He has published extensively on urbanisation, migration, and food security issues in the Global South. His most recent publications include *Rapid Urbanisation, Urban Food Deserts and Food Security in Africa* (with Jane Battersby), *Food and Nutrition Security in Southern African Cities* (with Bruce Frayne and Cameron McCordic), and a Special Issue of *International Migration* 55 (2017) on 'Cultivating the Migration–Food Security Nexus'.

Ann Donald is a researcher and teacher with a background in Geographical Information Systems. She was involved in the remote sensing and geographic analysis of the Consuming Urban Poverty project.

James Duminy is a researcher and PhD graduate at the African Centre for Cities (University of Cape Town) and general secretary of the Association of African Planning Schools. He holds Masters degrees in Town and Regional Planning (University of KwaZulu-Natal, South Africa) and Urban History (University of Leicester, United Kingdom) He completed his PhD at the University of Cape Town. His research interests centre on the interface between planning theory and history, with a focus on colonial and postcolonial Africa within the context of the wider Global South.

Issahaka Fuseini is a post-doctoral researcher at the African Centre for Cities (ACC), University of Cape Town (UCT). He is a human geographer with research interests in urban governance around the intersections of planning, provision and access to urban services, urban food security, and urban and peri-urban agriculture practices. He currently works on the Nourishing Spaces project, which is being implemented in six cities in Southern and East Africa; two each in Kenya, South Africa and Zimbabwe. The project focuses on understanding the urban food system changes and to mitigate the emergent burden of diet-related non-communicable diseases in urban Africa.

Patrick Odhiambo Hayombe is a senior lecturer and the dean of the School of Spatial Planning and Natural Resource Management at the Jaramogi Oginga Odinga University of Science and Technology (JOOUST). He received his PhD in Environmental Planning and Management and MPhil in Environmental Planning and Management at Moi University and his BA in Geography from the University of Nairobi. Dr. Hayombe has participated in several research projects, including Consuming Urban Poverty, Kisumu Local Interaction Platform – Mistra Urban Futures (SIDA), and Servir Africa/Regional Centre for Mapping and Resource Development, and supervised several post-graduate theses and published in refereed journals.

Gareth Haysom is a researcher at African Centre for Cities at the University of Cape Town where he coordinates the IDRC-funded Hungry Cities Partnership project. Gareth also works on the ESRC/DFID Consuming Urban Poverty project and the IDRC-funded Nourishing Space project. His work uses food as a lens to understand the complex nature of the urban transitions currently underway in the Global South.

Niraj Jain is a lecturer of land economy in the School of the Built Environment at the Copperbelt University (CBU) in Zambia. He holds an undergraduate degree from CBU and a MPhil from Cambridge University (United Kingdom). He is a Fellow of the Cambridge Commonwealth Society (FCCS) and of the Surveyors Institute of Zambia (FSIZ). Currently, Niraj sits on the Zambia Institute of Estate Agents Council, CBU Senate, and the national Higher Education Authority Committee of Experts. His research interests are mainly inclined towards land tenure security with an emphasis on customary land holding.

Samuel Kusangaya is a physical geographer with expertise in hydrology, Geographic Information Sciences (GIS), and remote sensing by profession and training. His interests are in geo-information science applications in climate change modelling, vegetation monitoring, and hydrology and water resource management. To date, Mr. Kusangaya has authored or co-authored more than 40 journal and reference papers, six course modules, four book chapters, and has developed and implemented various teaching curriculum. Currently Mr. Kusangaya is a lecturer within the Department of Geography and Environmental Science at the University of Zimbabwe and has submitted his PhD in Hydrology at the University of KwaZulu Natal, South Africa.

Oliver Manjengwa is a post-graduate student at the University of Cape Town where he is currently enrolled for a Masters in Southern Urbanism in the African Centre for Cities. He holds a BA in Geography from the University of Cape Town. Mr. Manjengwa has extensive research experience and has worked on several projects on urban poverty, urbanisation, risk and resilience in Zimbabwe.

Idah Mbengo is an environmental scientist in the department of Geography and Environmental Science at the University of Zimbabwe. She has over 15 years of teaching, research, and consulting experience. She specializes in environmental strategy formulation, environmental management systems, and social and environmental impact assessments. Her current research interests are in urban geography and cognitive and behavioural geography, with a particular focus the behavioural attitudes of smallholder farmers towards climate change.

Francis Muwowo is a lecturer in the Department of Urban and Regional Planning at Copperbelt University. He was previously the Town Planner of Lusaka. He holds a Masters in Town Planning from the University of Auckland and is the president of the African Council for Planning, Housing and Environment (ACOPHE).

Nelson Obange holds a PhD in Economics from Maseno University and is a faculty member in the Department of Economics at Maseno University. He has post-graduate training in project management and logistics. As a development economist and a researcher at KLIP, his focus is in urbanisation and urban economic challenges such as urban food insecurity and poverty, urban land use and governance, and urban public financing. He also has a passion for research on enhancing green energy and economies for better livelihoods.

Harun Okello Ogindo is currently an associate professor, researcher, and chair of the Department of Applied Plant Sciences at Maseno University, Kenya. He is a prominent researcher in the fields of agro-ecosystems, crop resource use, impacts of climate and environmental changes, and food systems. He worked for 15 years in implementing food security projects as an agronomist and extensionist within the Irrigation and Drainage Branch, Ministry of Agriculture, Kenya. He has spent a further 18 years teaching and researching on cropping systems of importance to food security among smallholder farmers in Eastern and Southern Africa.

Paul Otieno Opiyo is a programme management specialist with interests in food security, livelihoods, and environment. He has 20 years of experience managing development programmes in private, non-profit, and public sectors. He is currently a PhD candidate and researcher in urban food security based at Kisumu Local Interaction Platform Trust (KLIP). He holds a Bachelors of Environmental Science (Kenyatta University) and a MA in Project Planning & Management (University of Nairobi).

Fredrick Omondi Owino is a lecturer at the School of Spatial Planning and Natural Resource Management at Jaramogi Oginga Odinga University of Science and Technology (JOOUST). He received his Bachelors degree in Design and Masters degree in Urban and Regional Planning from the University of Nairobi. He completed a PhD in Planning from JOOUST after being awarded a scholarship by SIDA and Mistra Urban Futures (Kisumu

Local Interaction Platform). Currently he is a post-doctoral fellow whose research interests include green planning, human settlement planning, transportation, and urban agriculture. He has published in several refereed journals.

Liam Riley is a Banting Post-Doctoral Fellow at the Balsillie School of International Affairs in Canada. His contributions to urban food security research are centred on the use of a gender lens of analysis and on the implications of food insecurity for gender inequality in African cities. This work is rooted in extensive fieldwork in Malawi's major cities. He is currently managing a project on food security and food systems in secondary city case studies in Malawi (Mzuzu), Cameroon (Dschang), and Namibia (Oshakati) through the African Food Security Urban Network (AFSUN).

Muna Shifa is a post-doctoral research fellow, Southern Africa Labour and Development Research Unit, University of Cape Town. Her research is in development economics, with a particular focus on land tenure systems and rural livelihoods, subjective well-being and relative income, inequality and social cohesion, urbanisation and development, and analysis of poverty and inequality.

Lesley Sibanda is a researcher at the African Centre for Cities, University of Cape Town (UCT). She holds a BSc in Chemical Engineering and MPhil in Chemical Engineering, both from UCT.

Owen Sichone is the director of the Dag Hammarskjöld Institute for Peace and Conflict Studies, Copperbelt University. He has previously taught at several southern African universities including Wits University and the University of Cape Town in South Africa. He was also Nelson Mandela Professor in African Studies at Jawaharlal Nehru University in India. He is a political sociologist whose work includes studies of xenophobia and ethnic identities in South Africa and the state and democracy in Africa.

Caroline Skinner is the Urban Research Director at Women in Informal Employment: Globalizing and Organizing, and a senior researcher at the African Centre for Cities at the University of Cape Town.

Warren Smit is the research manager at the African Centre for Cities at the University of Cape Town, where he has worked since 2008. His background is in urban planning, and he previously worked as a researcher on urban issues from 1993 onwards (at the Urban Problems Research Unit, Built Environment Support Group, and Development Action Group). His research interests include urban governance, urban health, and housing.

Godfrey Tawodzera is an independent researcher, consultant, and academic based in Cape Town. He has previously worked at the University of

Limpopo, the University of Cape Town, the University of Zimbabwe, and the Catholic University of Mozambique. An associate of the African Centre for Cities (ACC), the Southern African Migration Programme (SAMP), and the African Food Security Urban Network (AFSUN), Godfrey researches on food security, poverty, livelihoods, and migration. He has published over 40 journals articles, book chapters, working papers, and monographs.

Percy Toriro is an urban planner who worked for many years in Harare City and taught at the University of Zimbabwe. He later worked at the Municipal Development Partnership for Eastern and Southern Africa, where he was an urban planning specialist managing the RUAF regional urban agriculture projects. He was later seconded to the AFSUN project. He is a part-president of the Zimbabwe Institute of Regional and Urban Planners. Percy's research interests include urban informality, urban environmental planning, and urban food security. He is currently pursuing PhD studies with the University of Cape Town.

George Godwin Wagah is a senior lecturer in the Department of Urban and Regional Planning and a former dean of the School of Planning and Architecture, Maseno University. He holds a PhD in Planning. He is the director of SPADE Project, a project implemented by Maseno University and Vrije Universiteit Amsterdam. He is also Principal Researcher of Socio Spatial Track Project, a project funded by Mistra Urban Futures. He is also a researcher with Consuming Urban Poverty, an ESRC/DFID-funded project. Wagah is a corporate member of the Kenya Institute of Planners (MKIP).

Preface

Food is central to the health and well-being of urban residents, and yet food is rarely on the urban agenda. This is particularly true in Africa, which is experiencing rapid urbanisation in both primary and secondary cities. The book argues that this lack of policy and programmatic attention has negative food security outcomes for urban residents, and therefore impedes economic and social development. It also makes a broader argument that important contributions to debates on urbanisation in sub-Saharan Africa, the nature of urban poverty and the relationship between governance, poverty and the spatial characteristics of cities and towns in the region can be made through a focus on urban food systems and the dynamics of urban food poverty.

The book emerges out of the ESRC/DFID-funded Governing Food Systems to Alleviate Poverty in Secondary Cities in Africa project (branded as Consuming Urban Poverty or CUP). The project, housed at the African Centre for Cities at the University of Cape Town (UCT), was a partnership between UCT, the Kisumu Local Interaction Platform (KLIP) in Kenya, the University of Zimbabwe in Zimbabwe, and Copperbelt University in Zambia. Working in Kisumu (Kenya), Epworth (Zimbabwe) and Kitwe (Zambia), the project sought to answer the central question of "How and why do the relational dynamics of urbanisation and the food system influence poverty dimensions and food security in secondary cities?" The chapters here represent some of the core findings and contributions of the project across its workstreams, which focused on: urbanisation and poverty, urban food systems and food security, and governance.

Carolyn Steel wrote, "In order to understand cities properly, we need to look at them through food" (Steel 2008: 10). We trust that this book enhances our collective understandings of African urbanism.

Jane Battersby and Vanessa Watson
University of Cape Town
March 2018

Reference

Steel, C. (2008). *Hungry city: How food shapes our lives*. London: Chatto and Windus.

Acknowledgements

We wish to acknowledge the support of the ESRC/DFID Joint Fund for Poverty Alleviation Research (grant number is ES/L008610/1) for the Governing Food Systems to Alleviate Poverty in Secondary Cities in Africa (Consuming Urban Poverty) project.

We would like to thank Carole Rakodi, chair of the Consuming Urban Poverty Advisory Board, and all the Board members for peer-reviewing the chapters and for valuable advice and guidance throughout the research project. We thank the full CUP research team, based in Cape Town, Kisumu, Epworth and Kitwe, for their contribution to the research effort. Also, Owen Crankshaw for advice on household surveys; Ivanna Katz for editing and formatting; Sean Robertson for graphics and Issahaka Fuseini for the case study map; and Tim Hardwick and Amy Johnston of Routledge for editorial assistance.

Introduction

Jane Battersby and Vanessa Watson

A call to connect food and urban agendas

Food is central to urban health, urban economies and urban form. Roberts has argued, "More than with any other of our biological needs, the choices we make around food affect the shape, style, pulse, smell, look, feel, health, economy, street life and infrastructure of the city" (2001: 4). However, food has been largely absent from urban studies and urban policy, and the urban has been equally absent from food security studies and policy. This book is the first of its kind that together brings the crucial issues of urban food security, food systems and governance. It highlights the implications of a lack of urban food policy and planning in African countries and cities. It also illustrates how household food security and food systems are impacted by governance decisions made in the absence of consideration of their potential food security impact. It is intended that the book provides new entry points to discuss the drivers of African urban development trajectories. It is based on the premise that important contributions to debates on urbanisation in sub-Saharan Africa, the nature of urban poverty, and the relationship between governance, poverty and the spatial characteristics of cities and towns in the region can be made through a focus on urban food systems and the dynamics of urban food poverty.

Food and poverty have historically been intertwined, with many national poverty lines constructed on the basis of the cost of a calorically sufficient diet. The Millennium Development Goals (MDGs) reinforced the notional connection between poverty and food security, with MDG 1 focusing on eradicating extreme poverty and hunger. However, while the ending extreme poverty component of MDG 1 was the focus of much policy attention, the hunger goal was termed the "forgotten MDG" by Graeme Wheeler, World Bank Managing Director, who argued that it was "often unrecognised, rarely acted upon, and grossly under-funded" (Fukuda-Parr 2017: 167).

Within the post-2015 Development Agenda, as characterized by the ratification of the Sustainable Development Goals (SDGs) and the New Urban Agenda (NUA), food security has risen in prominence. The second SDG is to: "End hunger, achieve food security and improved nutrition and promote sustainable agriculture," and within the NUA there are 18 mentions of food security or food more broadly, and a further seven mentions of nutrition. The framing of SDG 2 was not expressed as simply hunger, but also food security, and an important shift is its recognition of the need for improved nutrition in the context of sustainable agriculture. Food insecurity and hunger are not synonymous. Food security is the situation "when all people, at all times, have physical and economic access to sufficient safe and nutritious food that meets their dietary needs and food preferences for an active and healthy life" (FAO 1996). As will be elaborated on in Chapter 3 of this book, the achievement of food security is dependent on adequate availability, accessibility and capacity to utilize food, and the stability of this availability, accessibility and utilization. This widening of the agenda from ending hunger to achieving food security therefore requires a wider set of policy and programmatic responses.

While the increased prominence of food security in the global development agenda is welcomed, there are a number of concerns about the ways in which this development issue has been framed. SDG 2 speaks about addressing malnutrition in all of its forms, but there are no indicators within the monitoring framework to assess obesity (Battersby 2017b). It further fails to acknowledge the reality of *urban* food insecurity. Each of these is explained later.

In the transition from MDGs to SDGs one of the major shifts was the increased prominence of the issue of nutrition. Indeed, food security is increasingly termed "food and nutrition security," representing a shift in focus from caloric sufficiency (i.e. addressing hunger) to nutritional adequacy (i.e. addressing malnutrition), and therefore quality of diet. This shift is important as it adds complexity to the relationship between food and poverty, moving away from hunger being an indicator of poverty, to long-term malnutrition as being a driver of poverty.

The impacts of food and nutrition insecurity on human physical, social, emotional and cognitive development throughout the life course are well known (Pérez-Escamilla 2017: 1). However, increasingly arguments about the economic costs of food insecurity and malnutrition are being made in global development circles (IFPRI 2014). The corollary of this is that food and nutrition security is increasingly viewed as a critical entry point to alleviate various development challenges. The 2017 Global Nutrition Report, for example, argues that "a better nourished world is a better world" and that food and nutrition security play a catalytic role in achieving the SDGs (Development Initiatives 2017: 9).

However, in order for food and nutrition security to play this catalytic role, policies and programmes need to be contextually specific. Within SDG 2, the imagined context is of food and nutrition insecurity being located primarily amongst rural smallholder farmers. However, there is now a considerable evidence base that suggests food insecurity is increasingly an urban challenge in Africa. The 2017 Global Nutrition Report has urban food policies as one of its five key areas for action. The 2017 IFPRI Global Food Policy Report focused on Food Security and Nutrition in an Urbanising World (IFPRI 2017). Also in 2017, the Global Panel on Agriculture and Food Systems for Nutrition released their Urban Diets and Nutrition Policy Brief (Global Panel 2017). This renewed interest in urban food policy is the cumulative impact of research on urban food and nutrition security in the Global South.

During the last decade this has shifted from early efforts simply to get food insecurity in urban areas acknowledged by policy makers and funders, to a far more complex research agenda that situates household food insecurity within its wider food system context. There is also a new focus on changing retail systems (Battersby and Peyton 2014; Kroll 2016; Riley and Legwegoh 2014), urban infrastructure and its impact on the flows of food into and within the city (Wegerif 2017), and the resilience of urban food systems in the context of external shocks (Berger 2017). The recently published "Food and Nutrition Security in Southern African Cities" (Frayne *et al.* 2018) reflects this emerging research agenda.

Previous work on food security has conventionally focused at either the household scale or at aggregate food production, with far less focus on the food system itself and its intersection with cities. The High Level Panel of Experts on Food Security and Nutrition (HLPE) definition of a food system is: "A food system gathers all the elements (environment, people, inputs, processes, infrastructures, institutions, etc.) and activities that relate to the production, processing, distribution, preparation and consumption of food, and the outputs of these activities, including socioeconomic and environmental outcomes" (HLPE 2014: 12). This definition makes the connection between food security, the food system and the wider set of systems in which food operates. This systems approach therefore provides entry points to bring food into the urban agenda and *vice versa*.

Emerging findings from these various research projects, and those presented in this book, are at odds with the framing of the food system and food security as articulated in the SDGs and NUA. The framing of the food system within the SDGs remains predominantly focused on production and trade, with very little sense of the geographies of the food system and no focus on what happens within the city bounds, for example retail and consumption. The concept of a food system designed to meet the food security needs of urban residents in the NUA is informed by a normative vision of a territorial food system based on the concepts of the food shed, the bio region and place-making (Blay-Palmer *et al.* 2015). Here the food shed is the geographical location that produces food for a population, the bio region is a geographic region defined by the characteristics of the natural environment – rather than man-made divisions –, and place-making is a complex socio-political and spatial process that "underscores the importance of particular characteristics of the places in which food systems are evolving" (Blay-Palmer *et al.* 2015: 6).

The territorial food system position is most clearly articulated in Paragraph 95 of the NUA which commits to supporting "the implementation of integrated, polycentric and balanced territorial development policies and plans, encouraging cooperation and mutual support among different scales of cities and human settlements" (UN-Habitat 2016). While urban–rural linkages are undoubtedly important, the territorial or City Region Food System is a normative concept in danger of being treated by policy makers as objective reality. The call for food system planning at a territorial scale also provides a rationale for local government to neglect food system planning, particularly with regards to spatial planning, within municipal boundaries. It is also largely apolitical and ignores the power of multiple stakeholders, including large-scale private-sector actors, donor agencies, global trade agreements and other actors in shaping food system conditions. If effective food security interventions are to be developed, it is critical to understand the governance processes that shape the food system and the experience of food insecurity at global, national and municipal scales. This requires serious consideration of context, and of an empirically driven understanding of the local food systems and their intersections with urban form and function.

The work presented in this book suggests that the food systems of secondary cities in Africa are far more geographically diffuse than imagined within the SDGs and NUA. The structure of food systems and the food security outcomes that they create are shaped by land use planning decisions, infrastructure deficiencies, state responses to informality and a number of other specifically urban-scale factors. Given this, significant policy, planning and governance interventions are possible and required at the local government scale in order to address urban food security and poverty. The failure of the SDGs and NUA to call for specifically urban-scale policy allows for the continuation of a critical governance gap, in which food remains off the African urban planning and governance agendas (Battersby 2017b; Smit 2016).

This book therefore makes a series of contributions to the literature and emerging policy discourse on urban food security, food systems and urban governance through the findings from our case study work in Kisumu (Kenya), Kitwe (Zambia) and Epworth (Zimbabwe). It asks the question: what do the urban food systems in these secondary cities in Africa reveal about the dynamics of urban poverty and its governance? The next section of the Introduction highlights some of the innovative findings from our research on these three cities.

Innovative research findings

The first contribution is on the nature and scale of food insecurity in the case study cities researched in the Consuming Urban Poverty (CUP) project. Food insecurity levels in each case study city were high. Seventy-one percent of surveyed households in Kisumu were moderately or severely food insecure, 88% of the surveyed households in Epworth and 90% in the pro-poor sample in Kitwe. This supports other analyses of levels of food insecurity in African cities, including the pro-poor African Food Security Urban Network (AFSUN) surveys which, using the same food security indicator (Household Food Insecurity Access Prevalence – HFIAP), found high levels of food insecurity across a range of cities. The prevalence of the sampled households in poor neighbourhoods in Cape Town was 80%, 94% in Lusaka and 95% in Harare (Crush *et al.* 2012).

The CUP project used the suite of Food and Nutrition Technical Assistance (FANTA) food security indicators, as described in the next section. By using these indicators together it was possible to go beyond a headline food insecurity figure. By using the HFIAP in conjunction with the Months of Adequate Household Provision (MAHFP) and the Household Dietary Diversity Score (HDDS) it was possible to show that the sampled households generally had unexpectedly low numbers of months of inadequate household food provisioning in relation to the levels of food insecurity experienced. This relative stability of food access was secured by consistent access to a single staple food, maize. In Zambia in particular, agricultural and food security policy has been almost exclusively focused on ensuring access to maize. However, the low dietary diversity evident in the cities raises important questions about the nutritional adequacy of this kind of policy which views food security only as relief from starvation. The Lived Poverty Index (LPI) revealed the multidimensional nature of poverty in these cities, with households experiencing deficiencies in access to food, water, energy, medical resources and cash.

While food insecurity in cities is to a large degree an outcome of income poverty, our findings argue that it is also about the wider lived urban condition, in which inadequate access to water, sanitation and energy all shape households' food utilization capacity and therefore their food security. Our findings reinforce existing research on the compounding impact of multiple deprivations and, therefore, calls for greater acknowledgement of the multidimensional nature of poverty by policy makers and development agencies when developing poverty alleviation strategies.

The second innovative finding concerned the role of supermarkets in the food systems of these secondary cities. There is a considerable body of literature on the "supermarket transition" around the world. This work focusses largely on supply- and demand-side drivers of this retail transformation, and on the role of supermarkets in re-shaping supply chains in Africa. Our findings suggest that a broader conversation about supermarkets is necessary. Supermarkets intersect with food security, the food retail system, and urban governance in a number of ways that have not previously been the focus of significant research.

In terms of food security, in all case study cities (including Epworth, which had no supermarket within its bounds) poor urban residents were purchasing food from supermarkets. However, the main and most frequent sources of food remained markets and traders (mostly informal) within their immediate neighbourhoods. Supermarkets were far from the most important source of food. In Kisumu, by way of example, more than half of consumers bought less than 25% of their food at supermarkets. This suggests that although supermarkets are an important part of the food retail environment, purchasing patterns have maintained the historical dominance of frequent purchases from the local markets and less frequent large purchases from a more central location, be it the central market, a wholesaler or a supermarket.

While the supermarkets remain a minor source of food in terms of purchase, the findings of the project, in conjunction with the work of Abrahams (2009) and others, suggest that the supermarkets are changing the nature of the food retail sector. In the case of Kitwe, it was evident that a number of smaller traders were effectively using the supermarkets as their wholesaler. Items entering the city through the supermarket's supply chain are therefore resold in the informal sector. While supermarkets and local government problematized the location of street traders outside of supermarkets as parasitic and undermining the formal retail sector, our findings suggest a far more symbiotic set of relationships between the formal and informal sectors.

Further, much of the existing work on the expansion of supermarkets has focused on the various supply- and demand-side issues propelling the growth of the sector. There has been some focus on the role of improved urban infrastructure in enabling the expansion of supermarkets into new spaces. However, there has been little attention paid to the role of local government in supporting this expansion through granting planning permission. In Kisumu and Kitwe, shopping malls and the supermarkets that occupy them hold a position in the local government as symbols of modernization and economic opportunity. The findings here support work conducted in Cape Town which concluded:

> The expansion of malls and supermarkets must be understood in the context of four interconnected ideas currently shaping African cities, namely: the vision of the "modern" city; the imperative of economic growth; the notion of the rising middle class as the investment frontier; and the belief in public private partnerships as a means to achieve development objectives. (Battersby 2017a: 339)

There is far more research to be done on the drivers of food retail transformation in African cities.

The third contribution of the project is its mapping of the sources of food coming into the city, and the ways in which the findings from this work contribute to the emerging promotion of City Region Food Systems led by the FAO, RUAF Foundation, ICLEI and others. Within the City Region Food System literature a re-localized food system is seen as both the ideal and the current reality in smaller African cities (Jennings *et al.* 2015). However, the food systems for the three case study cities did not seem to correspond to this reality or hold this ideal as viable. As subsequent chapters will illustrate, these secondary cities were far more globally connected than the literature suggests, with fish in Kitwe and Kisumu travelling from China, eggs at the wholesale market in Kisumu coming from Uganda, and vegetables at the Chisokone Market in Kitwe coming from South Africa. These global connections were not limited to modern, formal supply chains, but were a central part of informal sector activities. Conversely, processed foods, such as *maputi* (Zimbabwean popcorn), were sometimes locally produced. The structure of the food system was determined by complex and interconnected factors beyond municipal, provincial or even national government competence. The sale of Ugandan over Kenyan eggs was the result of Ugandan agricultural policy which subsidized chicken feed as well as transport infrastructure investments that made cross-border trade along main arterial routes more viable than local trade along roads connecting Kisumu to neighbouring rural areas.

It is the contention of this book that the dominance of the City Region Food System as a policy framework promoted by the FAO and RUAF and in the NUA leads to a myopic focus on the immediate rural to urban linkages. In the analysis of food system governance within the case study cities it was apparent that the main point of policy and programmatic interventions for local governments was enhancing local production, with little understanding of the reasons why food was being sourced from beyond the city region. It was noted that many local production and processing projects had failed, but without knowing why. Within local governments there was no consideration of the role of large-scale, private-sector actors in shaping the food system, nor of the connections between formal and informal sector actors. The thinking about the food system was entirely focused on small-scale local actors. This suggests that the governance capacity and will to engage food systems planning is lacking, and that the call to re-localization from City Region Food System proponents may simply reinforce existing limitations to current food systems governance.

The final, and linked, set of findings is about the governance of urban food systems. The project found that there was generally weak governance of the urban food system with no clear mandate for local government and generally limited financial and personnel capacity within local government. This meant that many governance decisions that shaped the food system, such as decisions on traders, market sites and mall development, were made outside the context of a consideration of their food system impact. The relatively limited capacity of the state meant that key policy and planning decisions were strongly informed by national governments, large international donor agencies and private-sector actors. The book argues that these governance capacity issues are particularly evident in secondary cities, which are also the urban areas with the most pressing development challenges, vulnerable economic bases and worst access to up-to-date, accurate data on which to make governance decisions. These findings provide a useful entry point for research on wider urban development agendas and governance.

While the findings of this project speak explicitly to issues of food security and food systems, it is the belief of the project team that food provides an extremely useful lens through which to view urban poverty and governance. It provides us with a means to examine the complex connections between formal and informal sectors and the role of the state and other stakeholders in shaping the urban conditions that can alleviate urban poverty. The next section of the Introduction explains the conceptual framing and research methods behind the research in this book.

Urban poverty and urban food security: the research focus and methods

Research focus

The book is based on a three-year research project (funded by the ESRC and Dfid) known as CUP (Consuming Urban Poverty), located in the African Centre for Cities (ACC) at the University of Cape Town, and in collaboration with partner research teams at Copperbelt University (Zambia), the University of Zimbabwe and the Kisumu Local Interaction Platform (KLIP) – a research and policy knowledge hub facilitating urban research in Kisumu, Kenya. The full title of the project is *Governing Food Systems to Alleviate Poverty in Secondary Cities in Africa*. It builds on a longer trajectory of research within the ACC on urban food security in sub-Saharan Africa. This work started in 2008 with the establishment of AFSUN (the African Food Security Network), which was first hosted by Queens University and then by the Balsillie School of International Affairs (both in Canada), with the ACC as the lead developing partner institution.[1] AFSUN worked with eleven southern African research partners to explore the specificities of how poor African households experience urban food insecurity. CUP also collaborates with the Hungry Cities Partnership (HCP),[2] which researches urban food security across six Global South cities.

The CUP project argues that access to food and water are the most fundamental of human needs. In an urban setting, where inhabitants are usually disconnected from the direct production and distribution of food, inability to access food resources can be the most immediate and critical manifestation of the many dimensions of poverty. Yet policy interventions, particularly at the city scale, can make a significant difference to food security in contexts where even this most basic of human requirements is frequently lacking. The research therefore aimed to explore how urban poverty can be alleviated on the world's poorest continent through interventions in the urban food system.

Previous urban poverty inquiry (known as livelihoods research) focused at the household scale on specific determinants of deprivation (such as the lack of basic skills or access to sanitation) or governance, but the connections between this and the city scales of anti-poverty interventions in areas such as bulk infrastructure were not well conceptualized. Understandings of urban poverty as multi-dimensional emphasized the interconnected nature of natural, economic and governance systems which reinforce poverty, and this is still regarded as important (Mitlin and Satterthwaite 2013). Newer work on poverty reflects the discourse on social protection which emerged in the context of relatively robust growth of some middle-income nations (Mexico, Brazil, China), providing a belated recognition of the need for alternative institutional arrangements to protect the poverty-stricken and vulnerable against frequent livelihoods shocks (Cook and Kabeer 2009; Baulch 2011). Even more recently the conceptualization of social protection has moved beyond a narrow focus on risk management and safety nets to encompass a broad range of socio-economic policies including social security, health care, social insurance and child protection. This has been driven by the realization that the state needs to re-engage in the social arena, playing a more active role in shaping markets, redistributing gains from growth and ensuring adequate investments in the human capital and welfare of the poor (Ballard 2013).

However, cities and towns across the African continent vary significantly in size and nature. A great deal of urban research has tended to concentrate on the largest and capital cities of each country. But urbanisation in Africa is somewhat different as most growth is occurring in the intermediate and smaller cities: those with less than 500,000 people are now absorbing two-thirds of all urban population growth (UN-Habitat 2008a). There is little data to indicate the extent to which poverty and inequality vary between large and small cities, but given the generally lower capacity of local government outside the main cities as well as a tendency to concentrate national resources in infrastructure in the largest cities, it could be assumed that poverty is higher in smaller cities. Apart from a few exceptions (such as Namibia), smaller cities and towns in Africa also have higher rates of slum prevalence than capitals and large cities (UN-Habitat 2008b).

CUP therefore aimed to explore the cross-cutting theme of food and its links to poverty, governance and urban spaces. Taking the lens of the urban food system and the dynamics of food poverty allows an understanding of these various drivers of poverty, and provides scope for integrating multiple scales of analysis from the individual, household, neighbourhood, city and beyond. Understanding sources of food supply, global and local, has been an important element of this. Aspects of governance impact directly on food production, processing and distribution: urban laws and regulations governing use and access to land and to public spaces directly affect the food system, and particularly the spatial distribution of food retail. Likewise, access of the poor to services (especially water and energy) can support or hinder food production (Nissing and von Blottnitz 2010) whilst the presence or absence of other services in poor neighbourhoods (especially waste management) is a strong determinant of negative externalities of food in the urban system. The urban food system is a window onto the experience of poverty by urban residents and onto the impact of the dynamics of urban growth on the poor. The food system, one in which all urban residents have a stake, is also an analytical device that connects household livelihood strategies to global economic processes.

The case studies

Three cases of secondary cities (Kisumu, Kitwe and Epworth) in three African countries (Kenya, Zambia and Zimbabwe) were selected for research. The case study approach is ideally suited to in-depth analysis of complex urban conditions and processes in the African context (Duminy *et al.* 2014). It emphasizes the importance of context and is an approach that accords with recent calls for urban poverty research to understand how poverty dynamics and governance responses are shaped by contextual factors. The intention was not to claim that these cases are representative of secondary cities in Africa given the huge diversity of African urban settlements. Rather the aim was to maximize the theoretical potential of the cross-case analysis by minimizing broad contextual differences between individual cases. As such, the research design accords with the general principles of the "comparative case study of commonalities" (Ragin 1999). In this approach, the goal of the multi-case analysis is to "identify common causal conditions linked to a specific outcome across a relatively small number of purposefully selected cases" (Ragin 1999: 1141). The central question underlying the comparative case study of commonalities is *how* certain combinations of causal conditions produced the outcome observed in each case.

The selected cases are located in contexts that are culturally and historically similar, with British colonial legacies of urban administration and planning. They are located in low-income African countries and face acute problems relating to urban poverty and institutional governance. In terms of political stability, all three countries are democracies which regularly hold elections, but these are often marred by political violence and accusations of corruption and electoral manipulation. Development policy and governance dynamics in all three are characterized by a degree of anti-urbanism and a reluctance to accept or deal with problems of urbanisation and urban poverty, leaving poor urban residents vulnerable to economic and environmental shocks, including state-led evictions. As such, in broad terms the selected countries accord with the basic principles of "fragile states" with relatively weak state capacity to plan and manage urbanisation. Where the case studies do vary is in their role and function as towns: one is located within the peri-urban area of a primary city (Epworth near Harare in Zimbabwe), another is linked to mineral resource development (Kitwe in Zambia) and the third (Kisumu in Kenya) owes its existence and growth to its location as a transport junction.

Fieldwork

Four areas of fieldwork were carried out in each case study city. The first was a food retail mapping exercise and food retail survey. This involved the mapping of key retail sites, detailing the number and location of food retail outlets in particular areas across the city. It gave an understanding of the spatial distribution of food retail and how this might link to urban planning and governance, and to the location of transport nodes and business concentrations. From this mapping exercise, surveys were conducted with food retailers to gain an understanding of the kinds of foods being sold, the sources of the foods, common business practices and limitations experienced by retailers.

The second was a food pathways analysis involving an extended value chain analysis of five key food items per city. It tracked these foods from "fork to farm" using consumer choice of certain selected products and then tracing them back to their various supplier and producer sources, whether local or global. The foods tracked were specific to each city.

The third was a household food security survey. A limited number of in-depth household interviews were first carried out to help inform a wider survey and to gain an in-depth understanding of the daily food access strategies of the household. This was followed by a representative survey (varying by city) of 879 households in Kitwe, 841 in Kisumu and 483 in Epworth using a standardized household food security questionnaire. The core focus of the questionnaire was the FANTA food security indicators (HFIAP, HDDS, MAHFP) and the LPI. These indicators were used to describe household food poverty status in the studied neighbourhoods. The survey also collected data on where households sourced food from, as well as basic demographic data.

The fourth area of fieldwork was on processes and institutions of governance and was carried out through literature reviews and interviews with key informants in government and civil society.

Synthesizing these areas of work to address the over-arching research question, what do the urban food systems in these secondary cities in Africa reveal about the dynamics of urban poverty and its governance, has involved the testing of various concepts and theoretical positions on urban poverty against the data generated from the cases. Following the case method, it has been necessary to ask how the data speak back to current positions on urban poverty and food security.

The case study cities[3]

This section gives background to the three cities in which research was undertaken on governance (chapters 8–11), urban food systems (chapters 13–15), and urban food poverty (chapters 16–18) (Figure 0.1). Each of the city profiles below give a brief outline of settlement history, economy, population, governance and urban form.

Kisumu City, Kenya

Kisumu began as a regional hub for barter trade, earning its name, a Dholuo derivative, to describe a place to find food (UN-Habitat 2005). Located at the confluence of trading routes, Kisumu occupied a strategic trading point for the Maragoli, Nandi and Luo people long before the site was selected as a colonial commercial centre in 1898 (Nodalis Conseil 2013; UN-Habitat 2005). Noting its strategic location, the British colonial government established a railroad terminus and port at the existing settlement of Kisumu. This extended the reach of British colonial power, connected Kisumu more efficiently to the rest of the colony and facilitated trade in the entire Great Lakes region. Kenya won independence in 1963, and the following decades saw thriving sugar, cotton and fishing industries established in Kisumu. In the 1980s and 1990s Kisumu's economy stagnated and declined under economic liberalization (Aguilo *et al.* 2007). In the early 2000s, Kisumu began to develop rapidly, with bustling city streets and new construction projects (Geissler 2013). Its strategic location at the confluence of a major transport hub of the Great Lakes region linking Kenya, Uganda, Tanzania, Burundi and Rwanda makes it economically important beyond its local catchment area.

Growing from 53 km^2 in 1971, the municipal area of Kisumu now covers 260 km^2 of land area and 157 km^2 of water (UN-Habitat 2005). Kisumu's population has likewise expanded, from 32 431 shortly after independence in 1969, to 152 643 in 1979 and to 404 467 in 2009 (Mireri *et al.* 2007; Nodalis Conseil 2013). Today, Kisumu is home to a population of about 500 000 and had an estimated growth rate of 2.8% per year in 2006 (Geissler 2013; UN-Habitat 2006). It is the third largest city in Kenya (after Nairobi and Mombasa), and is one

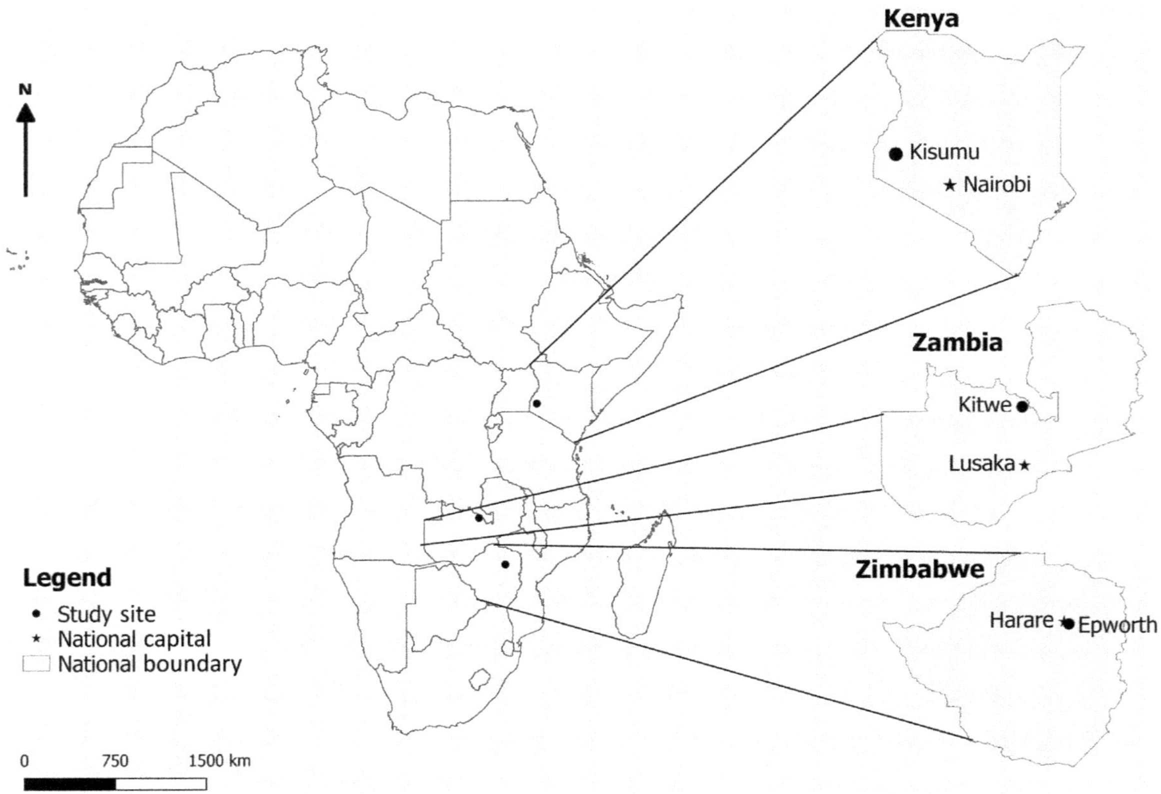

Figure 0.1 Location of case study cities

of Kenya's fastest growing cities (Gutberlet *et al.* 2017). Population figures are somewhat speculative, particularly since the redrawing of Kisumu's census areas.

Close to 50% of Kisumu residents live in absolute poverty, compared to the national average of 29% (UN-Habitat 2006). Unemployment in Kisumu is high, at 30% (Mireri *et al.* 2007; The World Bank 2016). The informal sector employs about 52% of the city's working population (or 70% in informal settlements) and supplies a monthly income of about KES 4000 (US$40) (Aguilo *et al.* 2007; Nodalis Conseil 2013; UN-Habitat 2005).

The average population density in Kisumu is 828 persons per km^2 (UN-Habitat 2006), but 7040 persons per km^2 in the informal settlements. It is unclear if these statistics are based on the entire municipal boundary, or solely the city's urban areas. Approximately 60% of Kisumu's population reside in informal settlements on just 8% of the city's land (Mireri *et al.* 2007). These areas have little in the way of infrastructure or services: one survey found that "85% of the plots lacked electricity, 83% lacked water on the plot and 51% lacked toilets" (Karanja 2010: 232). Most residents in informal settlements rent from landowners on an informal basis (Huchzermeyer 2009). The city's informal settlements are heterogeneous. Nyalenda has both low- and middle-income residents: upper Nyalenda is largely occupied by low-income households, while lower Nyalenda is largely occupied by middle-income households.

In terms of governance, Kisumu has become a sub-national administrative, commercial and communication centre for Western Kenya and was officially deemed a city under the Urban Areas and Cities Act, 2011. Kisumu City is now the capital of Kisumu County government. According to the Act, Kisumu should be governed by a municipal board in conjunction with a city manager. However, this board has never been established and instead the city is in the hands of an appointed interim city manager (Aoya 2016). Kisumu is the centre for Kenya's political opposition, and violent protests are common. There is also strong NGO and donor agency presence, many of them involved in urban governance and planning: in 2008 it was estimated that more than 900 NGOs and community-based groups operated within Kisumu City (Prince 2013).

The urban form of Kisumu has been strongly shaped by its colonial past. Following colonial urban planning policies, the city is roughly divided into concentric rings that mirror the city's original racially segregated zones. At the core is the old colonial centre, now the Central Business District (CBD), clustered around the old port (originally Block A); the CBD is surrounded by higher-class suburbs (formerly European and Asian residential areas, originally Blocks A and B), which in turn are surrounded by the city's "slum belt" (originally Block C and land excluded from the municipal boundaries) (UN-Habitat 2005; Karanja 2010). Peri-urban settlements and farmland extend beyond these informal settlements (Nodalis Conseil 2014). Low- and middle-income public housing is centred around government service stations in the city centre, although there has been no new investment in public housing since the 1980s (UN-Habitat 2005). Private estate developments are a growing housing option for the upper and middle classes (Nodalis Conseil 2013;

Apollo 2016). Most of these estates are located in the peri-urban northern and eastern outskirts of Kisumu, although a few are within the city's original upper-class Block B.

Most food is bought and sold in the city's many markets, which provide employment for an estimated 60% of the city's labour force (Nodalis Conseil 2013). Wholesale and retail markets cluster in busy city centre streets and in the peri-urban fringe (Nodalis Conseil 2013). Kisumu has an estimated 20 formal supermarkets, including the nation's leading supermarket brands of Uchumi and Ukwala. The role of Kisumu as a transport hub has a major impact on its food supplies, which are sourced both regionally and internationally. The city is positioned on the main highway between Busia (a prominent Uganda–Kenya border town) and Nairobi, while the Jomo Kenyatta highway serves as another main artery to connect Kisumu to Tanzania and surrounding towns and counties (CGoK 2013).

Kitwe, Zambia

Kitwe was founded in the 1930s as a mining town in the Zambia's Copperbelt, although the region was previously settled by the Lamba people (Siegel 1989). In 1924 Northern Rhodesia was declared a British Protectorate, and the colonial government began to remove the Copperbelt's established residents to Native Reserves to make way for the mines and farms staked out by the British South Africa Company (Siegel 1989). With the imposition of a tax that coerced the land's residents into a cash economy and necessitated their search for employment, a settlement of migrant mine workers emerged near the Nkana mine, creating the town of Kitwe (*The New York Times* 1987). The settlement grew quickly and in 1936 a Town Management Board was established; in 1953 Kitwe became a Municipal Council; and in 1967 Kitwe was officially declared a city (Hampwaye 2008).

Zambia's mines were privatized in the early 1990s, and Kitwe saw massive layoffs: from 1992 to 2002, 317 companies in Kitwe closed and more than 3000 workers were retrenched (Hampwaye 2008). By the early 2000s, 45% of Kitwe's labour force was unemployed (Smart *et al.* 2015). Since this time new non-mining investments have diversified the city's economy and Kitwe is now second only to Lusaka for its administrative and commercial importance. However, currently about 46% of household heads in Kitwe's informal settlements are employed in the informal sector, while only 35.4% have formal employment (Taylor *et al.* 2015). Income levels are low and more than 40% of those living in Kitwe's informal settlements live on less than US$40 per month (SDI n.d.).

Much of the town's resurgence appears to be fuelled by Chinese firms investing in mining and building hotels, malls, and stadiums in Kitwe and Ndola (Moriyasu 2016; Kayombo 2015; McGroarty and Parkinson 2016; Zambia Invest 2015). With a continued dependence on copper despite fluctuating copper prices, however, Kitwe's economy is still fragile (McGroarty and Parkinson 2016). The surrounding forests provide a source of timber and firewood but the area's soils are not ideal for intensive agricultural production (Sinkamba 2007; Hampwaye 2008).

The city occupies an area of 777 km^2 on a gently sloping plain. The population grew rapidly: at independence in 1964 it stood at 156 000. In the 1980s, depressed copper prices saw population growth slow in Kitwe, reaching just 347 024 people in 1990, and the privatization of the mines in the 1990s slowed growth further. By 2000, the city had grown to just 388 646 people (Hampwaye 2008) with the departure of many residents who migrated to Lusaka or to rural areas (Potts 2005). With reinvestment in the 2000s, Kitwe's population grew again: the 2010 census recorded a population of 517 543 people in Kitwe District, a 3.2% p.a. increase from the figure in 2000 (Central Statistical Office 2012). However, during the day, Kitwe swells to more than one million people with an influx of people, many working in the informal sector, from surrounding towns and rural areas (UN-Habitat 2009; Smart 2014).

In terms of governance, the Kitwe City Council (KCC) is responsible for making policy decisions for the city, providing services to the city's residents and collecting taxes (UN-Habitat 2009; Hampwaye 2008). However, its ability to adequately provide services to its constituents has been hampered by lack of funding from the national government, limited ability to collect taxes and an inefficient taxation system (Hampwaye 2008; UN-Habitat 2009).

Kitwe, like many cities colonized by the British, was built as a European "garden city," with spacious plots reserved for Europeans near the city centre, while Africans were relegated to locations on the city's outskirts (Nsemiwe 2006; UN-Habitat 2012). At independence in 1964, there were no informal settlements as the demand for housing was met by the mines' employee accommodation and by municipal housing (Tipple 1976; UN-Habitat 2012). With the growth of the population, the government embarked on projects to provide site-and-service housing, but the demand soon outstripped supply and informal settlements expanded rapidly. Today an estimated 50% of Kitwe's population reside in the city's 48 informal settlements (PPHPZ and ZHPPF 2014), all with very little service provision.

Spatially, Kitwe focusses on a central CBD with most activity around Chisokone Market, the city's main market that predominantly hosts small-scale businesses (Banda *et al.* 2004). The city centre also hosts a variety of retail outlets and service providers, such as the Shoprite food retail store, the Nkana Mall, the Kitwe Main Post Office and various banks, hotels and restaurants. The several new and large-scale shopping malls are dominated by South African brands. Meanwhile, the separate mining towns of Kalulushi and Chambishi, approximately 15 km west and 30 km north of Kitwe respectively, are growing centres with economic and social ties to Kitwe. The main Ndola–Kitwe dual carriageway (T3) connects Kitwe with the city of Ndola, about an hour's drive southeast of Kitwe, and with Chingola and other points north, including the Democratic Republic of Congo. These regional and cross-border links play an important role in Kitwe's urban food system.

Epworth, Zimbabwe

Epworth is a satellite town of Harare, located outside the city's municipal boundary. It was originally the site of two villages of subsistence farmers but under British colonial rule the land was claimed by the British South Africa Company. This land was donated to the Wesleyan Methodist Mission Trust in 1892 (Butcher 1988; Rakodi 1995). Epworth grew slowly over the years, at first with households associated with the mission station (Butcher 1986). During the peak of the Zimbabwean liberation war between the years 1966 and 1979 refugees fleeing conflict in the rural areas settled in Epworth as the area had no government-imposed movement or development restrictions, unlike other urban areas (Nyamvura and Brown 1999; Dialogue on Shelter 2009). In addition, Epworth's close proximity to the opportunities of Harare (then Salisbury) made it a logical destination for rural refugees (Butcher 1993; Dialogue on Shelter 2009). The church opened up the area to welcome refugees, resulting in the area's rapid growth in the late 1970s (Rakodi 1995). While the initial influx of refugees may have come from the surrounding areas and generally the Mashonaland area, later entrants came from across the country as the war intensified.

With independence in 1980, racial restrictions on movement to urban areas were lifted (Tibaijuka 2005). Epworth, with its lack of formal development, provided an easy destination for new arrivals, further increasing the settlement's population (Butcher 1988). This sudden growth in Epworth's population was beyond the administrative capacity of the Church, and the Methodist Missionary Trust donated part of their land to the government in 1986. The Epworth Local Board (ELB) was established as a local authority (Chirisa 2010) and it still governs Epworth today.

In Zimbabwe's depressed economy, the informal sector has become an important livelihood strategy in the face of growing unemployment. Informal employment is pervasive in Epworth, with only about 23% of the settlement's labour force employed in the formal sector (Mutisi 2014). Informal employment in Epworth is primarily in small and medium enterprises such as in carpentry, building, bicycle repair, welding and tin-smiting, motor mechanics, shoe making and repair, and retail. Some also participate in illegal sand extraction (Sigauke 2002). Any formal employment opportunities in Epworth are largely seasonal and are prone to the vagaries of political and economic instability. Those in the formal sector predominantly work as security guards, and as labourers in the nearby Msasa Industrial Site.

At independence in 1980 Epworth accommodated around 20 000 people (Butcher 1983; CSO 1982), and by the time of the 2012 census it had a population of 167 462 people (ZIMSTAT 2012). The population density of the settlement is about 4652 people per km^2. About 70% of these people live in the informal settlement locally known as "Magada," which is Ward 7 (Dialogue on Shelter 2009).

In terms of governance the Epworth Local Board has the tasks to administer and to regulate the growth of the settlement (Rakodi 1995), to oversee the area's development in terms of water, electricity and other services, and to spearhead its integration with the rest of Harare (Butcher 1998). The issue of housing in Epworth is highly political. While the ELB controls, at least in theory, the area's residential land allocation process, this is often not the reality. Since 2000, the ruling ZANU PF party has guaranteed the security of tenure of thousands of households in Epworth under the condition of their political affiliation (Marongwe *et al.* 2011). This has been done outside of the Board's relevant channels, creating planning challenges for the Board and further constraining its capacity to deliver housing and other services.

About 70% of the residents live in informal conditions (ZIMSTAT 2012). Housing typologies in Epworth (Makunde 2016; Dialogue on Shelter 2009) range from permanent housing built from brick and mortar with an asbestos or corrugated iron roof (generally found in older and more planned residential areas), to informal housing in Ward 7, where structures are generally made from poles, mud or sunburnt bricks (Chirisa 2010). Most homeowners have no title deeds, electricity is found in only a few areas and there is limited access to tap water (Dialogue on Shelter 2009). Public facilities and services are limited, but there are several schools and two clinics. There are no supermarkets (although there was one briefly in a failed mini-mall development on the edge of Epworth) and most food and household goods are sold through informal traders. There are eight areas in Epworth zoned as commercial shopping precincts in which vending is permitted. Two main tarmacked roads, Chiremba Road and Deport Road, connect Epworth with Harare, which is some 20km away.

Structure of the book

The book is divided into four parts. This section of the Introduction explains the division and summarizes the chapters in each of these parts.

The overview of the research which informed this book (discussed earlier) established its aim as exploring the connections between urban poverty, urban food systems, urban household food security and urban governance in three case study cities. Analyzing these dimensions of the food problem required fieldwork in the cities but was also dependent on data from continent-wide, national and municipal sources. The book is therefore divided as follows.

Part I deals with the over-arching aspects of urbanisation and urban poverty in sub-Saharan Africa and the current arguments and positions within the field of urban food security which link these aspects together. Part II deals with urban food governance and planning. It highlights the importance of understanding policy shifts historically, the role which contemporary forms of governance play in relation to food and how governance influences the food system in the three case study cities. Part III covers urban food systems. Chapters highlight the importance of value chain analysis, and how food systems operate in the three case study cities drawing in part on value chain work in these contexts. Part IV deals with urban food poverty. The three chapters draw on quantitative and qualitative data-gathering in the three cities to explain how poor households are affected by and navigate the issue of food security. The next section explains the content of these four parts of the book in more detail.

Part I: urbanisation, poverty, food and measurement

These chapters provide an overview of the wider debates on urban food security in Africa. A key argument of the book is that there is a link between urban food insecurity and urban poverty in sub-Saharan Africa: that food insecurity is more severe in cities and towns due to the nature of urbanisation in Africa and the ways in which institutional governance and policy often functions to worsen the problem. Chapter 1 sets out the debates over the relationship between urbanisation and development in Africa, suggesting that the usual assumption that growing cities drive development through economic growth does not necessarily hold in Africa. There is evidence that urban growth occurs without economic growth, and jobless growth (along with increasing poverty) is taking place in many parts of the continent, although some authors contest this. The high levels of urban food insecurity which accompany poverty have been worsened by decades of international food security policy which has been dominated by rural bias, a focus on rural hunger and supporting smallholder agriculture (Chapter 2). Urban food security has therefore been largely ignored in policy terms and rural bias continues despite some recognition of its presence in the NUA.

Chapter 3 argues that where research has been conducted on urban food security it has been misdirected, focusing attention entirely at the household scale and neglecting the systemic drivers of food insecurity which take place through the wider food and urban systems. This neglect has also resulted in inappropriate and ineffective policies and programmes and has hence done very little to address urban food security issues. However, attempts to measure some of these underlying dynamics in African cities (urbanisation, poverty, inequality and labour markets) have proved extremely challenging. Chapter 4 describes some of the major data gaps on the continent and the kinds of problems which researchers face when attempting to acquire data. This is why any large-scale data-driven conclusions have to be treated with a degree of caution.

Part II: urban food governance and planning

This section of the book reviews the relationship between urban governance in African cities and its influence on urban poverty, urban food systems and urban food security. This supports a key argument in the book that changes to urban programmes, policies and regulations at various levels of government, and particularly in urban governance, can have a significant impact on improving urban food security and poverty.

Chapter 5 provides a historical overview of how the colonial Kenyan state engaged with urban food problems, to support the idea that little attention has been given to the local governance and planning of urban food systems. It also shows how current rural bias in food policy has its roots in colonial approaches to food, in the case of Kenya, which understood the food "problem" in particular ways. Chapter 6 explains contemporary food governance and planning systems on the continent including the very wide range of actors beyond government which have an influence on the urban food system. The chapter shows the importance of understanding this range of actors and how they interact with each other in particular contexts in order to develop processes and programmes to address urban food issues.

As the case study chapters on food systems and food security show, the informal economy (or sector)[4] plays a central role in urban food systems and urban food security. Chapter 7 explains the importance of informal suppliers and retailers in shaping access by poor households to nutritious and fresh food, as well as their role as a source of employment and income, particularly for women. The dynamics of informal food trade are also being shaped in new ways as a result of growing supermarket penetration, with the relationship between them varying between displacement and co-existence.

Chapter 8 discusses the planning and governance of food systems in Kisumu (Kenya), where the food system is shaped by forces that extend well beyond the city, the region and even the continent. Yet local government in Kisumu has no policies addressing food security, and there is no recognition of how the existing plans, policies and laws relating to urban planning and management affect urban food security, often negatively.

Chapter 9 discusses the planning and governance of food in Kitwe (Zambia) where the food system also extends well beyond the boundaries of the city and region. As in Kisumu, local government continually intervenes in aspects of the city through actions which impact on the management of space (especially market retail spaces). Many local government interventions are also driven by a normative view of the idealized "modern" city in which informality in all its forms is viewed as undesirable and in need of eradication. These direct and indirect governance actions all impact on urban food insecurity, almost always in negative ways.

Chapter 10 discusses the planning and governance of food in Epworth (Zimbabwe), where, as elsewhere, the food system plays itself out from global to local scales. This is particularly the case in towns in Zimbabwe as the national government has frequently introduced new laws and financial restrictions which influence food supplies. The chapter discusses the interrelated system of formal and informal food networks and rule-making shaped by policies, plans and instruments of all kinds, most of which negatively impact on household food security.

Chapter 11 discusses urban agriculture in Harare (including Epworth) as a great deal of "productionist" food policy suggests that this can provide a solution to the problem of food security. The chapter shows that despite the apparent policy support for this practice, urban agriculture policy has been ambiguous, and while there is a degree of state tolerance of the practice, city plans and planning officials tend to wish it away. And while urban agriculture is practiced, food security remains extremely high.

Part III: understanding the urban food systems

This section of the book explains the operation of the food system in the three case study cities. Central to understanding this concept of a food system is the idea of food value chains. Chapter 12 explains food value chain analysis from a lifecycle perspective, based on the three cases. Following selected food items, the chapter tracks the foods from source, through transport, sale and processing to final waste disposal. It analyzes significant environmental and social impacts which undermine sustainability and food security, and identifies "hotpots" in the chain which threaten health and safety.

Chapters 13, 14 and 15 present the findings of the analysis of the food system in Kisumu, Kitwe and Epworth, respectively. The research documented the distribution and nature of formal and informal retail outlets, and in Kisumu and Epworth reverse value chain analysis tracked the supply of five key food items. Both value chain analyses dispelled the common assumption that towns primarily source their food from their adjacent rural regions. In Kisumu most food supplies come from the rest of Kenya, other African countries and globally. Epworth, as well, has relied on both imported and local food sources largely due to national policy interventions.

Research on food distribution networks shows the dominance of the informal sector in food retail and especially women traders. It also shows how informal traders cluster in poorer neighbourhoods, at transport terminals and sometimes outside formal supermarkets. In Epworth research showed the significant impact of the political and economic turmoil which has affected Zimbabwe over the last decades, leading to periods of food shortage and price spirals. State currency and banking interventions have led to frequent shifts in the way food is imported and sold. Low investor confidence has slowed formal foreign and local supermarket growth and hence food supply is even more dependent on the informal sector than elsewhere.

Part IV: the state of urban food poverty and its connections to the food system

This section of the book shows the importance of understanding the multi-dimensional nature of urban poverty and how access to food plays a role in this. The chapter findings are based on quantitative surveys of households and a limited number of in-depth household interviews in neighbourhoods in each city. These findings show the lived realities of food poverty experienced by poor households and how they negotiate food access over time. All three cases suggest extremely high levels of food insecurity in these three secondary cities.

Chapter 16 presents the findings of food insecurity research in Kisumu. The results show high levels of food insecurity with 71% of households being either moderately or severely food insecure according to the HFIAP measure. While this level is high, it is better than the situation in sampled areas of Kitwe and Epworth. High levels of food insecurity together with low dietary diversity constrain the economic and productive potential of the population and increase the risk of ill-health and poverty. As in Kitwe and Epworth (the next two chapters), households used a range of strategies to manage their food supply, with supermarkets being used less frequently than other kinds of retail outlets.

Chapter 17 presents the findings of food insecurity research in two poor neighbourhoods of Kitwe. The results show high levels of food insecurity with 90% of households experiencing either moderate or severe food insecurity. The research shows that food poverty among the urban poor in Kitwe is due to general income poverty underpinned by structural macroeconomic factors beyond the control of the household, although household-scale factors such as size and structure contribute to the magnitude of the phenomenon. The ability of urban households to access quality food regularly in the right quantities transcends mere availability of food to encompass other structural systems such as the broader food system and urban system, socio-cultural variables and non-food factors such as the spatial characteristics of the urban settlement and governance interventions.

Chapter 18 presents the findings of food insecurity research in Epworth. The results show acute food insecurity with 88% of surveyed households being moderately or severely food insecure. Households experienced high levels of financial stress as a result of unemployment, low income and rising costs of living. As a result, most households were suffering from insufficient and non-nutritious food with low levels of dietary diversity, all of which have major health implications. The research also shows how households navigate the urban food system in order to maximize their food security through changes in where they access food, the kinds of foods they purchase and how they negotiate aspects such as credit.

Together these chapters illustrate the complex interconnections between lived poverty, food systems and governance systems. They suggest that the food security is a critical challenge facing secondary cities in Africa, and that the state is currently poorly capacitated or informed to act. It is intended that the cumulative outcome of this book is a case for greater attention to be paid by urban scholars to food as means to understand and alleviate poverty.

Notes

1 AFSUN www.afsun.org/
2 HCP http://hungrycities.net/
3 These profiles draw on unpublished project city profiles prepared by Rebekah Adams for Kisumu and Kitwe, and Rebekah Adams and Godfrey Tawodzera for Epworth.
4 In this book we use the term "informal sector" based on the 2003 definitional norms of the International Conference of Labour Statisticians (ICLS). Vanek *et al.* (2014:6) summarize these norms:

> The "*informal sector*" refers to unincorporated enterprises that may also be unregistered and/or small; "*informal employment*" refers to employment without social protection through work both inside and outside the informal sector; and the "*informal economy*" refers to all units, activities, and workers so defined and the output from them.

References

Abrahams, C. (2009). Transforming the region: Supermarkets and the local food economy. *African Affairs*, 109(434), pp. 115–134.

Aguilo, P., L'Esperance, A., Mbau, E., Palmer, P., Patel, A. and Sparkman, T. (2007). *Attracting investment to Kisumu: Opportunities and challenges*. New York, NY: Columbia University School of International and Public Affairs.

Aoya, C. (2016). Senator calls for establishment of Kisumu city board. *Hivi Sasa*. Available at: www.hivisasa.com/kisumu/politics/115971 [Accessed 15 Jul. 2016].

Apollo, S. (2016). No housing for low-income earners in Kisumu as developers focus on high-end clients. *Daily Nation*. Available at: www.nation.co.ke/lifestyle/DN2/No-housing-for-low-income-earners-in-Kisumu/-/957860/3032564/-/jmx3pz/-/index.html [Accessed 14 Jul. 2016].

Ballard, R. (2013). Geographies of development II: Cash transfers and the reinvention of development for the poor. *Progress in Human Geography*, 37(6), pp. 811–821.

Banda, C., Mutula, S. and Grand, B. (2004). Information needs assessment for small scale business activity in Zambia: Case study of Chisokone Market in Kitwe. *Malaysian Journal of Library & Information Science*, 9(2), pp. 95–108.

Battersby, J. (2017a). Food system transformation in the absence of food system planning: The case of supermarket and shopping mall retail expansion in Cape Town, South Africa. *Built Environment*, 43(3), pp. 417–430.

Battersby, J. (2017b). MDGs to SDGs—new goals, same gaps: The continued absence of urban food security in the post-2015 global development agenda. *African Geographical Review*, 36(1), pp. 115–129.

Battersby, J. and Peyton, S. (2014). The geography of supermarkets in Cape Town: Supermarket expansion and food access. *Urban Forum*, 25(2), pp. 153–164.

Baulch, B., ed., (2011). *Why poverty persists: Poverty dynamics in Asia and Africa*. Cheltenham: Edward Elgar Publishing.

Berger, M.T. (2017). *Rethinking resilient urban food systems. Vulnerability to food insecurity as a consequence of drought—the case of Cape Town*. Unpublished Masters, Utrecht University.

Blay-Palmer, A., Renting, H. and Dubbeling, M. (2015). *City-region food systems: A literature review*. Leusden: RUAF Foundation. Available at: www.ruaf.org/sites/default/files/City%20Region%20Food%20Systems%20literature%20review.pdf [Accessed 21 Feb. 2018].

Butcher, C. (1986). *Low income housing in Zimbabwe: A case study of the Epworth squatter upgrading programme*. RUP Occasional Paper No. 6, University of Zimbabwe.

Butcher, C. (1988). *Epworth: A review of the densification policy*. Report Prepared for the Epworth Local Board.

Butcher, C. (1993). Urban low-income housing: A Case Study of the Epworth Squatter Settlement Upgrading Programme. In L. Zinyama, D. S. Tevera, and S. D. Cumming, eds., *Harare: the growth and problems of the city*. Harare: University of Zimbabwe Publications, pp. 61–76.

Central Statistical Office (CSO), (2012). *2010 census of population and housing report: Population summary report*. Lusaka: Central Statistical Office.

Chirisa, I. (2010). An analysis of the environmental stewardship concept and its applicability in peri-urban towns: Lessons from Epworth in Zimbabwe. *Journal of Sustainable Development in Africa*, 12(4)(Summer B Issue.), pp. 41–57.

Cook, S. and Kabeer, N. (2009). *Socio-economic security over the life-course: A global review of social protection*. Prepared as the final report of a Social Protection Scoping Study funded by the Ford Foundation, Centre for Social Protection, Institute of Development Studies, Sussex.

Crush, J., Frayne, B. and Pendleton, W. (2012). The crisis of food insecurity in African cities. *Journal of Hunger & Environmental Nutrition*, 7(2–3), pp. 271–292.

CSO, (1982). *Census national report*. Harare: CSO.

CGoK (County Government of Kisumu), (2013). *Kisumu county: First county integrated development plan 2013 - 2017*. Kisumu.

Development Initiatives, (2017). *Global nutrition report 2017: Nourishing the SDGs*. Bristol: Development Initiatives.

Dialogue on Shelter for the Homeless in Zimbabwe Trust, (2009). *Epworth profiling report*. Harare: DSHZT.

Duminy, J., Andreasen, J., Lerise, F., Odendaal, N. and Watson, V., eds., (2014). Planning and the case study method in Africa: The planner in dirty shoes. Basingstoke: Palgrave Macmillan.

FAO, (1996). *Declaration on world food security*. Rome: FAO World Food Summit. Available at: www.fao.org/docrep/003/w3613e/w3613e00.htm [Accessed 21 Feb. 2018].

Frayne, B., Crush, J. and McCordic, C., eds., (2018). *Food and nutrition security in Southern African cities*. London: Routledge.

Fukuda-Parr, S. (2017). *Millennium development goals: Ideas, interests and influence*. London: Taylor and Francis.

Geissler, P. (2013). Stuck in ruins, or up and coming? The shifting geography of urban public health research in Kisumu, Kenya. *Africa*, 83(04), pp. 539–560.

Global Panel, (2017). *Urban diets and nutrition: Trends, challenges and opportunities for policy action. Policy Brief, No. 9*. London: GlobalPanel on Agriculture and Food Systems for Nutrition.

Gutberlet, J., Kain, J.H., Nyakinya, B., Oloko, M., Zapata, P. and Zapata Campos, M.J. (2017). Bridging weak links of solid waste management in informal settlements. *The Journal of Environment & Development*, 26(1), pp. 106–131.

Hampwaye, G. (2008). *Decentralization, local economic development, and urban agriculture in Zambia*. PhD, University of Witwatersrand.

HLPE (High Level Panel of Experts), (2014). *Food losses and waste in the context of sustainable food systems*. A report by the High-Level Panel of Experts on Food Security and Nutrition of the Committee on World Food Security. Rome: HLPE.

Huchzermeyer, M. (2009). Enumeration as a grassroot tool towards securing tenure in slums: Insights from Kisumu, Kenya. *Urban Forum*, 20(3), pp. 271–292.

IFPRI (International Food Policy Research Institute), (2014). *Global nutrition report 2014: Actions and accountability to accelerate the world's progress on nutrition*. Washington, DC: IFPRI.

IFPRI (International Food Policy Research Institute) (2017). *Global food policy report, 2017*. Washington, DC: IFPRI.

Jennings, S., Cottee, J., Curtis, T. and Miller, S. (2015). *Food in an urbanised world: The role of city region food systems in resilience and sustainable development*. London: International Sustainability Unit. Available at: www.fao.org/fileadmin/templates/agphome/documents/horticulture/crfs/foodurba nized.pdf [Accessed 22 Feb. 2018].

Karanja, I. (2010). An enumeration and mapping of informal settlements in Kisumu, Kenya, implemented by their inhabitants. *Environment and Urbanisation*, 22(1), pp. 217–239.

Kayombo, M. (2015). Chinese firm to invest $50m in Dola Hill mall. *Zambia Daily Mail*. Available at: www.daily-mail.co.zm/?p=46922 [Accessed 14 Oct. 2016].

Kroll, F. (2016). *Deflating the fallacy of food deserts: Local food geographies in orange farm and inner-city Johannesburg*. Working Paper 38. Cape Town: PLAAS, UWC and Centre of Excellence on Food Security.

Makunde, G. (2016). Challenges in urban development control and housing provision: a case of Epworth, Chitungwiza and Harare, Zimbabwe. *International Journal of Technology and Management*, 1(1), pp. 1–12.

Marongwe, L., Mukoto, S. and Chatiza, K. (2011). *Scoping study: Governance of urban land markets in Zimbabwe.* Urban Landmark: Pretoria.

McGroarty, P. and Parkinson, J. (2016). Mining collapse cripples Africa's dreams of prosperity. *The Wall Street Journal.* Available at: www.wsj.com/articles/mining-collapse-cripples-africas-dreams-of-prosperity-1457104328 [Accessed 5 Oct. 2016].

Mireri, C., Atekyereza, P., Kyessi, A. and Mushi, N. (2007). Environmental risks of urban agriculture in the Lake Victoria drainage basin: A case of Kisumu municipality, Kenya. *Habitat International*, 31(3–4), pp. 375–386.

Mitlin, D. and Satterthwaite, D. (2013). *Urban poverty in the global south.* London and New York, NY: Routledge.

Moriyasu, K. (2016). Chinese companies and Africa need each other like never before. *Financial Times.* Available at: www.ft.com/content/636015a8-5a5f-11e6-8d05-4eaa66292c32 [Accessed 14 Oct. 2016].

Mutisi, L. (2014). *An investigation into the contribution of housing developments to wetland degradation within the city of Harare, Zimbabwe.* Unpublished Masters Thesis, University of South Africa: Pretoria.

New York Times, (1987) Glory that was copper lets Zambia down, *New York Times*, 17 June 1987, p. 18. Available at https://www.nytimes.com/1987/06/17/world/glory-that-was-copper-lets-zambia-down.html [Accessed 25 April 2018].

Nissing, C. and von Blottnitz, H. (2010). An economic model for energisation and its integration into the urban energy planning process. *Energy Policy*, 38, pp. 2370–2378.

Nodalis Conseil, (2013). *Kisumu ISUD-Plan.* Paris: Nodalis.

Nodalis Conseil, (2014). *Kisumu ISUD-Plan, Part 2: Planning for Kisumu's future. The plan. January 2014.* Paris: Nodalis.

Nsemiwe, N. (2006). Negotiating the interface: Struggles involved in the upgrading of informal settlements—a case study of Nkandabwe in Kitwe, Zambia. In: *Shaping the change XXIII FIG Congress, Munich, Germany.* Available at: www.fig.net/pub/fig2006/papers/ts27/ts27_03_nsemiwe_0737.pdf [Accessed 1 Mar. 2018].

Nyamvura, T. and Brown, A. (1999). Epworth—an informal, unplanned and squatter settlement: Upgrading Initiatives in Epworth. In: A. Brown and C. Davidson, eds., *Responsive design and plan implementation.* Cardiff: Department of City and Regional Planning, Cardiff University, pp. 107–109.

Pérez-Escamilla, R. (2017). Food security and the 2015–2030 sustainable development goals: From human to planetary health: Perspectives and opinions. *Current Developments in Nutrition*, 1(7), p. e000513.

Potts, D. (2005). Counter-urbanisation on the Zambian Copperbelt? Interpretations and implications. *Urban Studies*, 42(4), pp. 583–609.

PPHPZ (People's Process on Housing and Poverty in Zambia) and ZHPPF (Zambia Homeless and Poor People's Federation), (2014). *Building city-wide sanitation strategies from the bottom up: A situational analysis for Kitwe, Zambia.* Available at: www.gov.uk/dfid-research-outputs/building-city-wide-sanitation-strategies-from-the-bottom-up-a-situational-analysis-for-kitwe-zambia [Accessed 7 Oct. 2016].

Prince, R. (2013). 'Tarmacking' in the millennium city: Spatial and temporal trajectories of empowerment and development in Kisumu, Kenya. *Africa*, 83(4), pp. 582–605.

Ragin, C. (1999). The distinctiveness of case-oriented research. *HSR: Health Services Research*, 34(5), pp. 1137–1151.

Rakodi, C. (1995). The household strategies of the urban poor: Coping with poverty and recession in Gweru, Zimbabwe. *Habitat International*, 19(4), pp. 447–471.

Riley, L. and Legwegoh, A. (2014). Comparative urban food geographies in Blantyre and Gaborone. *African Geographical Review*, 33(1), pp. 52–66.

Roberts, W. (2001). The way to a city's heart is through its stomach putting food security on the urban planning menu. *Crackerbarrel Philosophy Series*. Toronto: Toronto Food Policy Council. Available at: https://s3.amazonaws.com/academia.edu.documents/30355006/tfpc_hs_report.pdf?AWSAccessKeyId=AKIAIWOWYYGZ2Y53UL3A&Expires=1524658694&Signature=EnBbLShzd5LCMyX5McjDLcNXngA%3D&response-content-disposition=inline%3B%20filename%3DThe_Way_to_a_Citys_Heart_is_Through_Its.pdf [Accessed 15 Feb. 2018].

SDI, (n.d.). Water and Sanitation Action Research in the City of Kitwe: Community mapping towards inclusive development. Policy brief. *Shack/Slum Dwellers International*. Available at: https://assets.publishing.service.gov.uk/media/57a089e640f0b649740002fa/Kitwe_Zambia_POLICY_BRIEF.pdf [Accessed 7 Oct. 2016].

Siegel, B. (1989). The "wild" and "lazy" Lamba: Ethnic stereotypes on the Central African Copperbelt. *Anthropology Publications*, Paper 5. Available at: http://scholarexchange.furman.edu/ant-publications/5 [Accessed 6 Sept. 2016].

Sigauke, N. (2002). *Improving urban livelihoods: Epworth sustainable livelihoods benchmark study*. Harare: ITDG Southern Africa.

Sinkamba, P. (2007). *Technical and financial proposal for sustainability of the Copperbelt Environment Project in Zambia*. MPhil, University of Stellenbosch.

Smart, J. (2014). *Urban agriculture and economic change in the Zambia Copperbelt: The cases of Ndola, Kitwe and Luanshya*. PhD, University of Otago.

Smart, J., Nel, E. and Binns, T. (2015). Economic crisis and food security in Africa: Exploring the significance of urban agriculture in Zambia's Copperbelt province. *Geoforum*, 65, pp. 37–45.

Smit, W. (2016). Urban governance and urban food systems in Africa: Examining the linkages. *Cities*, 58, pp. 80–86.

Taylor, T., Banda-Thole, C. and Mwanangombe, S. (2015). Characteristics of house ownership and tenancy status in informal settlements in the City of Kitwe in Zambia. *American Journal of Sociological Research*, 5(2), pp. 30–44.

The World Bank, (2016). *Republic of Kenya: Kenya urbanisation review. Report No: AUS8099*. Washington, DC: The World Bank.

Tibaijuka, A. (2005). *Report of the fact-finding mission to assess the scope and impact of 'Operation Murambatsvina' in Zimbabwe*. UN-Habitat Special Envoy on Human Settlements, 18 July 2005, UN Habitat Nairobi.

Tipple, A. (1976). The low-cost housing market in Kitwe, Zambia. *Land Development Policies*, 41(244), pp. 148–152.

UN-Habitat, (2005). *Situation analysis of informal settlements in Kisumu*. Cities without slums sub-regional programme for eastern and southern Africa. Nairobi: United Nations Human Settlements Programme.

UN-Habitat, (2006). *Kisumu urban sector profile*. United Nations Human Settlements Programme. Available at: http://unhabitat.org/books/kenya-kisumu-urban-profile/ [Accessed 5 Jul. 2016].

UN-Habitat, (2008a). *The State of African Cities 2008: A framework for addressing urban challenges in Africa*. Nairobi: UN-Habitat.

UN-Habitat, (2008b). *State of the World's Cities 2008/2009: Harmonious Cities*. London: Earthscan.

UN-Habitat, (2009). *Zambia: Kitwe urban profile*. Nairobi: UN-Habitat.

UN-Habitat, (2012). *Zambia urban housing sector profile*. Nairobi: United Nations Human Settlements Programme.

UN-Habitat, (2016). *New Urban Agenda*. Quito: United Nations. Available at: http://habitat3.org/wp-content/uploads/NUA-English.pdf [Accessed 21 Feb. 2018].

United Nations, (2014). *The World Urbanisation prospects: The 2014 Review* (ET/ESA/SER.A/352). UN Department of Economic and Social Affairs, Population Division.

Vanek, J., Chen, M., Carré, F., Heintz, J. and Hussmanns, R. (2014). Statistics on the informal economy: Definitions, regional estimates and challenges. *WIEGO, Working Paper (Statistics) 2*. WIEGO, Cambridge, MA.

Wegerif, M.C. (2017). *Feeding Dar es Salaam: A symbiotic food system perspective*. PhD, Wageningen University.

Zambia Invest, (2015). USD 3 million Chinese investment opens up hotel in Kitwe, Zambia. *Zambiainvest.com*. Available at: www.zambiainvest.com/tourism/usd-3-million-chinese-investment-opens-hotel-in-kitwe-zambia [Accessed 14 Oct. 2016].

ZIMSTAT, (2012). *Zimbabwe population census 2012 Harare provincial report*. Harare: ZIMSTAT.

Part I

Urbanisation, poverty, food and measurement

1 African urbanisation and poverty

Muna Shifa and Jacqueline Borel-Saladin

Introduction

Current trends and projections of urban growth have led to increased recognition of the importance of cities. The New Urban Agenda (NUA) refers to urbanisation as "one of the 21st century's most transformative trends" (United Nations General Assembly 2016: 3). Notably, the Agenda recognises the unique urban developmental challenges in developing and least developed countries – especially in Africa – and calls for special attention to be paid to them. This emphasis on African cities seems necessary, given the longstanding perception that Africa is experiencing an "historically rapid rate of urbanisation" (AfDB/OECD/UNDP 2016). By some estimates, the share of urban residents grew from 14% in 1950 to 40% today, with the figure expected to reach 50% by mid-2030. Despite this, both the extent of urbanisation and the relationship between urbanisation and development in Africa are still hotly contested. In this chapter, we will discuss how and why this contestation has developed. We will begin by discussing how development, economic growth, and urbanisation are defined and then consider the relationship between them. This then leads to a discussion of the conflicting evidence for this relationship in Africa. The existing body of research on this relationship has focused predominantly on large African cities. This, however, means that there is a knowledge gap regarding smaller African cities, which is concerning, considering their growing importance as sites of urbanisation in Africa. It is therefore for this reason that the research of the Consuming Urban Poverty (CUP) project focuses on food security and poverty specifically in the context of secondary African cities.

Defining development and urbanisation

The term "development" is often assumed to refer to economic development only; however, broader conceptualisations of the term consider more than just economic growth as measured by Gross Domestic Product (GDP) (Scutaru 2013). These variables include unemployment, poverty, education, health, environment, migration, overcrowding, and inequality. One of the best known of these more comprehensive indicators is perhaps the Human Development Index (HDI), which measures health by life expectancy at birth, education by mean of number of years of formal schooling, and standard of living by Gross National Income (GNI) per capita. When focusing on the economic growth aspect of development, it is important to remember that simply increasing GDP or GNI does not necessarily lead to development. Inclusive growth is required, that being a long-term, sustainable approach to economic growth, with equal opportunity for a large part of the labour force to access markets and resources (Ianchovichina and Lundstrom 2009). This type of economic growth is essential to poverty reduction.

Defining urbanisation is complicated by several factors. In sub-Saharan Africa (SSA), different criteria are used for defining urban areas, ranging from population size to the main economic activity and government function (Cohen 2004; McGranahan and Satterthwaite 2014). Differing types of boundaries for an urban area further confuse the issue (for a discussion on data-related issues, see Chapter 4 and Borel-Saladin (2017a, 2017b)). Some researchers argue that more accurate measurement can be achieved by focusing on the economics of urbanisation (Potts 2017). Taking this approach, the main marker of urbanisation would be a change in economic activity, i.e. residents no longer predominantly engaged in agricultural activity to survive but employed in industry and tertiary sectors. However, some argue it is not feasible to add economic criteria to the definition of "urban" currently employed in most of Africa for several reasons (Turok 2017), one being that economic data at the sub-national scale is even less reliable and more patchy than population data. Complicating the definition of urban areas with an economic dimension would therefore jeopardise further the already infrequent and uneven collection of important basic demographic information. In addition, the data collected for the newly defined urban areas would be difficult to compare to historical data due to lack of consistent information for earlier periods.

Another argument regarding the definition of urbanisation proposes that the spatial-demographic conceptualisation of urbanisation cannot simply be ignored in favour of an economic conceptualisation as large, concentrated groups of people have different needs to rural areas regardless of the dominant type of economic activity taking place, e.g. different infrastructure (Fox *et al.* 2017). In addition, it is suggested that using the traditional multiple census and survey data sources, combined with remote sensing and migration studies over long periods, can help refine the accuracy of urban population estimates. There are also some promising developments using geospatial datasets which combine remotely sensed data and census data with innovative modelling techniques, e.g. the Global Rural–Urban Mapping Project (GRUMP http://sedac.ciesin.columbia.edu/data/collection/grump-v1), the Atlas of Urban Expansion (Angel 2012), WorldPop (www.worldpop.org.uk/), and Africapolis (www.oecd.org/swac/topics/africapolis/).

The terms "urbanisation" and "urban growth" are also often used interchangeably. These terms are not synonymous, however, as "urbanisation" refers to an increasing proportion of the total population living in urban areas, while "urban growth" refers to the absolute growth of the urban population. The perceived interchangeability of these terms is thus particularly problematic when referring to SSA where urban population growth rates are high but urbanisation rates are historically low (Fox *et al.* 2017). "Urban expansion" (which refers to the physical expansion of urban areas) is another term often used to mean increased urbanisation, yet the physical area occupied by an urban settlement may increase without population growth (resulting in declining population densities), or the population may increase without an increase in the size of the area occupied (resulting in increased densities).

Several scholars argue that natural population increase rather than net rural–urban migration might be the major driving force for the observed rapid urban growth in Africa (Kessides 2007; Potts 2012b; Parnell and Pieterse 2014; Fox *et al.* 2017). Many African cities have in the past, and continue today, to grow quickly, with population growth rates faster than those of European cities during the height of their expansion in the 19th Century (Potts 2012a). However, technically speaking, for urbanisation to occur, the urban population growth rate must exceed the national growth rate, and in African cities most urban population growth patterns generally tend to follow national population growth patterns. Nevertheless, the populations of many African countries have been growing fast (Potts 2012a). Improved technology and disease control resulting in access to surplus food supplies and reduced mortality are believed to have contributed significantly to this rapid population growth in African cities since 1960 (Fox 2012). High fertility and low mortality rates in rural areas have also resulted in rural settlements reaching the population threshold to be reclassified as urban areas (Fox *et al.* 2017). Growing populations in rural areas with limited resources may also create a push to urban areas.

That being said, each African city still has its own unique history and path of development. This includes migration patterns that are complex and responsive to multiple factors, with people migrating for a variety of different reasons over time (Oyeniyi 2013). Understanding and measuring these migration flows is obviously important for accurately gauging urban population size and change; however, the longstanding assumption of sustained, uniformly high levels of rural–urban migration across the African continent is increasingly being challenged. The reclassification of rural areas as urban areas has, for example, probably contributed more to urbanisation in SSA than it has been credited with, and rural–urban migration less so (Fox *et al.* 2017). Thus, continued high levels of urban population growth and the complications this entails remain a very real issue in SSA.

Urbanisation, economic growth and poverty reduction

One of the ways cities are considered to drive development is through economic growth, yet this view takes for granted the built-in positive relationship between urbanisation and growth (Turok and McGranahan 2013). Instead, it is the specific form of urbanisation that determines if it enables growth. As such, in some cases the concentration of people and businesses in cities can facilitate economic growth by increasing productivity through the effects of specialisation and scale (Collier 2016), which in turn can allow increased efficiency from large-scale, specialised production, the reduction of transaction costs, and the sharing of knowledge among firms – the so-called agglomeration economies (Turok and McGranahan 2013). However, this pooling of people and business activities in cities does not necessarily guarantee increased economic growth.

Although economic growth and urbanisation are positively related in the context of other developing regions, the evidence is mixed in the case of Africa. Several studies have found no positive relationship between urbanisation and economic growth or development in Africa (Bouare 2006; Bloom and Khanna 2007; Ravallion *et al.* 2007; Brückner 2012; Turok 2013; Gollin *et al.* 2016). Bouare (2006) even found a negative correlation between urbanisation and GDP for 23 of 32 SSA countries between 1985 and 2000. Likewise, Turok (2013) found no relationship between urbanisation levels and levels of GDP per capita for selected African countries over the period 1985–2010. However, in contrast, using the HDI as a measure of development, Njoh (2003) found a positive relationship between urbanisation and development in 40 SSA countries. Likewise, Kessides (2007) found a positive relationship between urbanisation and economic growth in 15 of the 24 countries considered between 1990 and 2003. However, the same study showed a negative relationship between GDP and urbanisation for the remaining nine countries. Potts (2016) argues that though some of the findings of these studies could be due to significant variations in annual growth of GDP per capita in most African countries in the 1980s and 1990s, that could not be adequately captured by the few data points used. The significant fluctuations in GDP per capita growth in SSA can be seen in Figure 1.1, which presents growth in levels of urbanisation and GDP per capita in SSA based on five-year intervals.

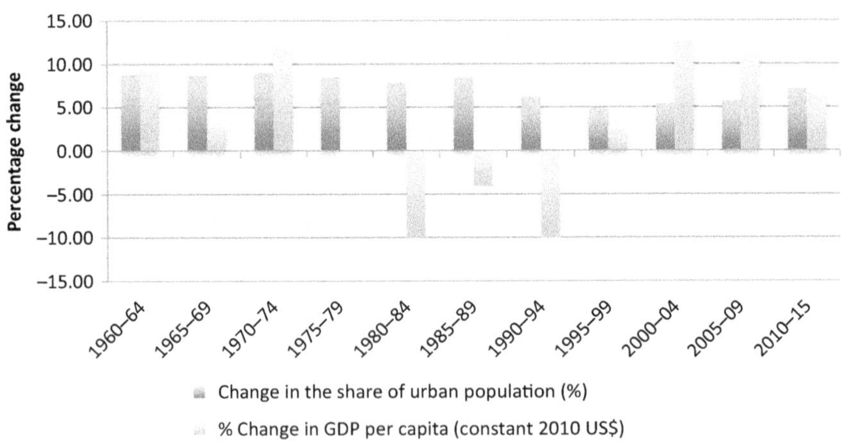

Figure 1.1 Percentage changes in the share of population and GDP per capita in SSA between 1960 and 2016

(Authors' calculations using data from World Bank development indicators)

Figure 1.1 shows that SSA countries registered higher economic growth over the period 1960–1975. Rapid urbanisation also occurred in this period across the continent. This was mostly a response to the lifting of influx control measures imposed by colonial governments, which had previously limited migration to many African cities (Satterthwaite 2007). Strong economic growth from the 1960s and concomitant investment in, and development of, urban areas also supported increased rural–urban migration (Potts 2012a). Economic growth during this period was driven mainly by both high resource exports and structural transformation (Jedwab 2011; De Vries *et al.* 2015). For instance, using data from 11 countries, a recent study showed that manufacturing's employment share grew from 4.7% in 1960 to 7.8% in 1975 (De Vries *et al.* 2015). These structural changes, which resulted in the re-allocation of workers from agriculture to urban sector jobs in manufacturing and service sectors also contributed to the increase in rate of urban growth. Thus, although part of the initial increase in urban growth in Africa was due to the relaxation of migration laws after independence in many countries, evidence suggests that economic growth and the associated structural transformation also played a role (Weeks 1994).

After 1975, however, there is disagreement about the rate of urbanisation in SSA countries (Potts 2006, 2009, 2012b; Satterthwaite 2010). This is because the lack of up-to-date, reliable census data led to international organisations and researchers relying on population predictions based on the earlier period of rapid urbanisation. Concurrently, SSA countries registered slow or negative economic growth over the period 1975–1995, attributable to various economic and political crises (e.g. oil crises, ethnic conflict, and war) and the regulations around structural adjustment programmes to which many African countries were subjected. The result was the widely held belief that SSA countries continued to urbanise rapidly throughout the 1980s and 1990s despite low levels of economic growth (Satterthwaite 2010). This has been held up as a stark contrast to the positive association between urbanisation and economic growth found in most developed countries, leading to the long-held belief that urbanisation occurred without economic growth in Africa (United Nations 2015).

As stated, however, this premise has been challenged with the argument that urbanisation in most of SSA was either very slow or stagnated in the 1980s/1990s (Kessides 2007; Potts 2006, 2009, 2012b; Satterthwaite 2010). Potts (2008) argues that rural–urban migration decreased in the face of less spending on urban areas, due to the oil crisis and structural adjustment programmes that led to slowing economic growth in this period. Figure 1.1 confirms this, as it illustrates how urbanisation also started slowing down after 1975 and declined significantly in the 1990s. Increasing circular migration and shorter periods of time spent in towns countered the flow of rural–urban migrants leading to a decline in urbanisation in SSA following the weakening of African economies since the 1980s. As Potts (2012b: 14) states, "confronted by economic insecurity and other hardships worse than where they came from, people behave as rationally in Africa as anywhere else", and left the urban areas.

Although economic growth in SSA started to recover after 1995, the percentage increase in urbanisation increased modestly. In addition to natural urban growth, oil discovery in several SSA countries may have also contributed to the increase in the level of urbanisation in recent years (Potts 2013). More recent data show that some African countries experienced increased urbanisation with GDP growth from 2003 to 2014, thanks to strengthening commodity prices (Potts 2017). Still, continued high rates of urbanisation have been incorrectly assumed in some countries. There are also variations in the level of urbanisation in different regions of Africa over time. For instance, looking at the countries in the CUP project, the urban share of the population increased far more in Zambia than in Kenya and Zimbabwe between 1950 and 1980 (Figure 1.2). Zambia's urban population increased by more than 200% during this period, and then declined by 12% between 1980 and 2000. The comparison of urbanisation and per capita GDP over time in Figure 1.3 shows that this de-urbanisation occurred mainly in response to economic decline after the 1980s (Potts 2006). A similar pattern of decline in urban population share and GDP is observed in Zimbabwe since 2000. Zimbabwe, subject to continuous, long-term political conflict and its negative economic consequences, was the only one of the three countries in the CUP study to experience a drop in the proportion of its urban population between 2000 and 2015. There is a general upward trend in the share of the urban population in Kenya, but a decline in the rate of growth post 1980, mirroring the lower annual GDP growth rates from that point onwards. Thus, the notion that SSA countries urbanised without economic growth is misleading (Kessides 2007; Annez and Buckley 2009).

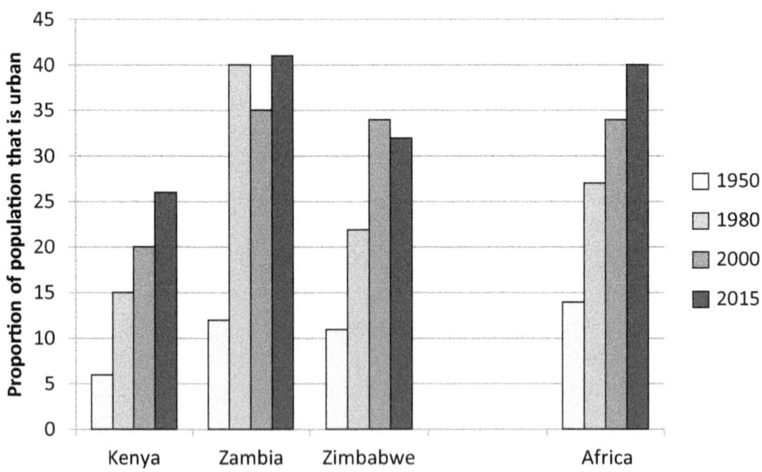

Figure 1.2 Urban population proportion dynamics in Kenya, Zambia, and Zimbabwe, 1950–2015

(World Urbanisation Prospects: The 2014 Review)

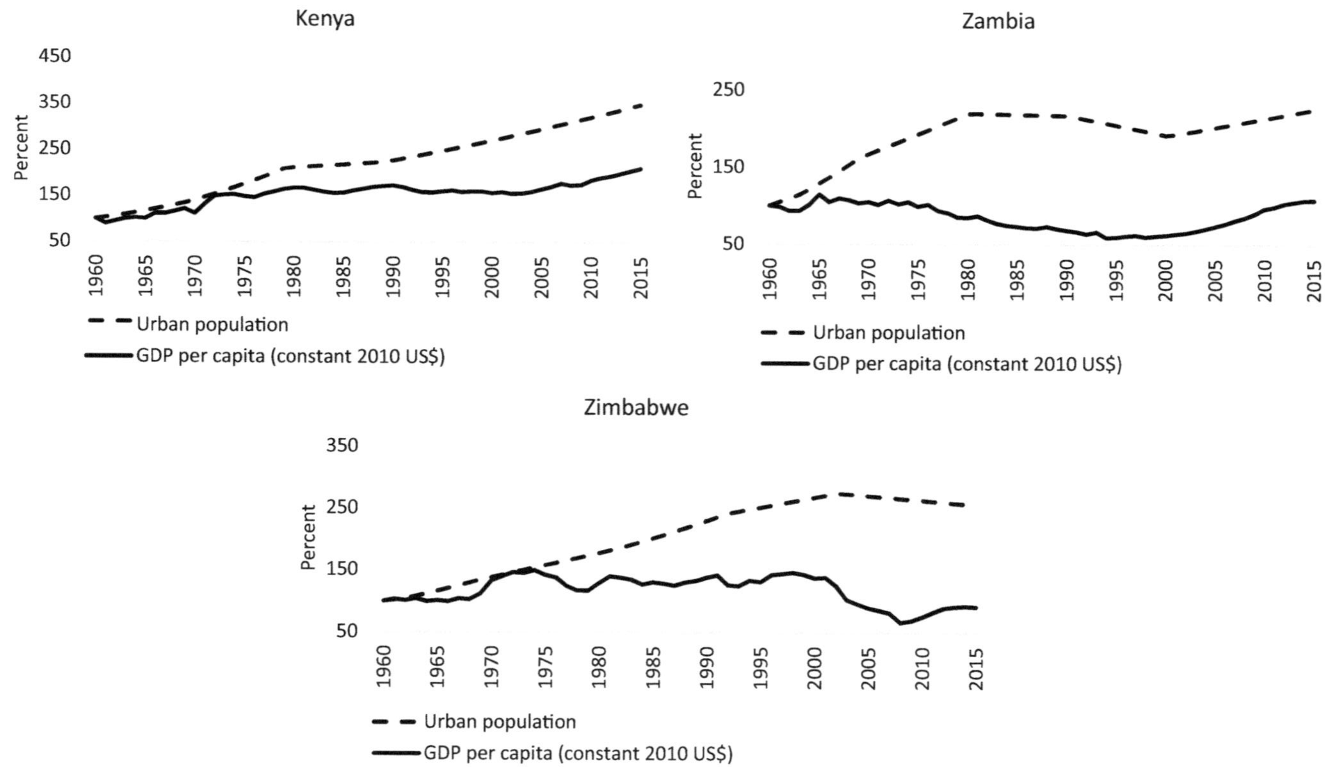

Figure 1.3 Trends in urbanisation and per capita GDP in Kenya, Zambia, and Zimbabwe

(Authors' calculations based on World Bank development indicators)

(Note: GDP is measured at 2010 constant (purchasing power parity) international dollars)

As Kessides states: "Africa cannot simply be characterised as 'urbanisation without growth', and the term does not even fit many of the countries" (Kessides 2007: 481).

Where there is economic growth, the extent to which it reduces poverty depends on how the growth pattern affects income distribution, which in turn depends on several other factors (Ferreira *et al.* 2010). SSA has one of the highest levels of income inequality, which has been increasing since the early 1990s (Lakner and Milanovic 2015). In addition, inequalities in income or other living standard outcomes are higher in urban areas compared to rural areas (Deaton and Dreze 2002; Sahn and Stifel 2003). These high levels of inequality can result in a weakened or non-existent links between economic growth and poverty reduction. This implies that the key problem is that most countries in Africa experienced rapid urban growth with lower levels of "inclusive" economic growth (Kessides 2007; Annez and Buckley 2009).

The link between urbanisation and poverty reduction can also be affected by the nature of SSA's structural transformation since the 1990s. In the case of SSA, evidence suggests that although the share of agriculture started to drop slowly after the 1990s, more workers were allocated to non-tradable service sectors and the employment share in industry hardly rose (De Vries *et al.* 2015; Gollin *et al.* 2016). Unlike the manufacturing sector, re-allocation of workers to non-tradable services is not necessarily related to increased aggregate productivity and poverty reduction in SSA (De Vries *et al.* 2015; Gollin *et al.* 2016). As a result, despite most of the countries in SSA recording positive economic growth rates since 2000, the extent of poverty reduction has been comparatively low in comparison to other developing regions. The percentage of the population living below the US$1.90/day income poverty line declined from 57% in 1990 to 50% in 2005 and 43% in 2012. In contrast, the absolute number of people living below this poverty level increased from 288 million in 1990 to 389 million in 2012 (The World Bank 2013), while during the same period the number of people living below US$1.90 dollar a day decreased by 847 million (85%) in East Asia and the Pacific (EAP), and it decreased by 265 million (46%) in South Asia (SAS). Although there is no question that rural poverty rates exceed urban poverty rates in SSA, urban poverty is also increasing in the region. The poor are also urbanising faster, leading to the "urbanisation of poverty" (Ravallion *et al.* 2007: 8; Mitlin and Satterthwaite 2013). According to income poverty estimates from Ravallion *et al.* (2007), for urban areas in SSA using a US$1.08/day international poverty line, the share of the urban population living under US$1.08 a day increased from 24.3% in 1993 to 30.2% in 2002 in SSA.

Secondary cities

Not all cities are the same. This is as true of African cities as those in other parts of the world: "Africa is not all one thing, and no one country or city can be used as the example to stand in for all the rest" (Myers 2011: 104). This thus suggests that the extent of both urban growth and poverty are not uniformly distributed across different urban centres (Cohen 2006; Ferré *et al.* 2012). This is as true for small urban centres as it is for large ones. The development of small cities is in fact crucial for several reasons. The combined population of the world's small cities is very substantial, as half of all urban inhabitants in 2014 lived in smaller cities of under 500 000 people (United Nations 2015). It is predicted that most of the world's urban population will be living in smaller cities under one million residents by 2030. Moreover, most of the fastest growing cities in the world have fewer than one million residents and are in Africa and Asia. Many migrants will migrate first to a smaller urban area, as it is closer in proximity to their rural home. These places thus "facilitate navigation of the non-farm economy . . . allowing a broader base of the poor population to become physically, economically, and socially mobile" (Ingelaere *et al.* 2018: 1). Thus, secondary cities have the potential to facilitate inclusive growth and poverty reduction in urbanising developing countries. The dynamics of urbanisation and poverty are, however, particularly poorly understood outside of larger and capital cities in Africa. This is due to the difficulty of obtaining disaggregated data at the smaller city level, and it means that most discussions around poverty focus on the gap between urban and rural areas. There are also marked differences in levels of poverty and access to services between different sized cities. The existing few studies suggest that small cities tend to suffer from much poorer service delivery than larger cities, as well as higher rates of poverty and infant and child mortality (Cohen 2006; Ferré *et al.* 2012; Shifa and Leibbrandt 2017a). For example, Shifa and Leibbrandt (2017a) found that multidimensional poverty estimates varied greatly across the different hierarchies of cities and towns in Kenya. The four dimensions they used for this measure were: education, health, living standards, and asset holdings (see Shifa and Leibbrandt (2017a) for details on the approach used to calculate the multidimensional poverty estimates). The percentage of population who were living below the multidimensional poverty line was found to be relatively lower in the largest city, Nairobi (27%), and its satellite towns, Ruiru (22%) and Thika (27%). The poverty rate was relatively higher in other secondary cities such as Mombasa (44%), Kisumu (46%), and Garissa (61%). In addition, a further disaggregated poverty analysis also indicated that there are large spatial disparities in poverty levels between different locations within large and secondary cities including Nairobi, Mombasa, and Kisumu. These findings suggest that comparing living standards across different urban centres based on average welfare estimates hides significant differences between and within urban centre inequalities.

Likewise, using census data from Zambia, Shifa and Leibbrandt (2017b) found that although the proportion of people who are deprived of access to basic services, such as access to safe drinking water and improved sanitation, is still higher in rural districts or smaller towns, these figures are increasing faster in large urban centres such as Lusaka, Ndola, and Livingstone. Compared to 1990 estimates, the extent of deprivation of safe drinking water in 2010 increased by 189 percentage points in Lusaka district, 212 percentage points in

Livingstone, and 308 percentage points in Ndola. These results suggest that the extent of provision of basic services and well-paying job opportunities is not sufficient to accommodate the rapid urban growth in the capital or smaller cities. Such challenges will likely be even greater in smaller cities since their population growth tends to be faster than larger cities. Also, smaller cities often lack the institutional capacity at the local government level to effectively manage this growth and cope with the associated challenges. Given the relatively poor understanding of urbanisation and urban growth in smaller African cities, what these processes mean for development and poverty reduction is still unclear.

Conclusion

Starting from a low level of urbanisation, African countries have experienced a significant increase in urbanisation over the past five decades. Many argue that earlier urban growth was linked to rapid natural urban population increase; however, without effective industrial development since the 1980s, a large portion of the urban population continues to work in the informal sector. Investment and solid urban policy are needed to help grow the economies of cities to ensure that their residents have a better standard of living, including access to sufficient services and livelihood opportunities in healthy environments. Economic growth alone is not a sufficient condition to address urban poverty though, as higher inequality stifles the poverty-reducing effect of economic growth. Another important aspect of Africa's urban transition is that a large and potentially growing share of the urban population lives in intermediate and smaller towns, where investment in infrastructure and job creation is very limited. Tracking this urban transition and understanding its associated challenges are crucial for effective governance that facilitates sustainable cities in Africa.

References

AfDB/OECD/UNDP, (2016). *African economic outlook 2016: Sustainable cities and structural transformation*. Paris: OECD Publishing.

Angel, S. (2012). *Planet of cities*. Cambridge, MA: Lincoln Institute of Land Policy.

Annez, P.C. and Buckley, R. (2009). Urbanisation and growth: Setting the context. In: M. Spence, P.C. Annez, and R.M. Buckley, eds., *Urbanisation and growth. Commission on growth and development*. Washington, DC: The World Bank, pp. 1–45.

Bloom, D.E. and Khanna, T. (2007). The urban revolution. *Finance and Development*, 44(3), pp. 9–14.

Borel-Saladin, J. (2017a). Where to draw the line: Data problems and other difficulties estimating urbanisation in Africa. *Development Southern Africa*, 34(2), pp. 137–150.

Borel-Saladin, J. (2017b). Data dilemmas: Availability, access and applicability for analysis in sub-Saharan African cities. *Urban Forum*, 28(4), pp. 333–343.

Bouare, O. (2006). Levels of urbanisation in Anglophone, Lusophone and Francophone African countries. In: C. Cross, D. Gelderblom, N. Roux, J. Mafukidze, eds., *Views onmigration in sub-Saharan Africa: Proceedings of an African migration alliance workshop*. Cape Town: HSRC Press, pp. 48–73.

Brückner, M. (2012). Economic growth, size of the agricultural sector, and urbanisation in Africa. *Journal of Urban Economics*, 71(1), pp. 26–36.

Cohen, B. (2004). Urban growth in developing countries: A review of current trends and a caution regarding existing forecasts. *World Development*, 32(1), pp. 23–51.

Cohen, B. (2006). Urbanisation in developing countries: Current trends, future projections, and key challenges for sustainability. *Technology in Society*, 28(1), pp. 63–80.

Collier, P. (2016). African urbanisation: An analytic policy guide. *Oxford Review of Economic Policy*, 33(3), pp. 405–437.

Deaton, A. and Dreze, J. (2002). Poverty and inequality in India: A re-examination. *Economic and Political Weekly*, 2002, pp. 3729–3748.

De Vries, G., Timmer, M. and de Vries, K. (2015). Structural transformation in Africa: Static gains, dynamic losses. *The Journal of Development Studies*, 51(6), pp. 674–688.

Ferré, C., Ferreira, F. and Lanjouw, P. (2012). Is there a metropolitan bias? The relationship between poverty and city size in a selection of developing countries. *The World Bank Economic Review*, 26(3), pp. 351–382.

Ferreira, F.H., Leite, P.G. and Ravallion, M. (2010). Poverty reduction without economic growth? Explaining Brazil's poverty dynamics, 1985–2004. *Journal of Development Economics*, 93(1), pp. 20–36.

Fox, S. (2012). Urbanisation as a global historical process: Theory and evidence from sub-Saharan Africa. *Population and Development Review*, 38(2), pp. 285–310.

Fox, S., Bloch, R. and Monroy, J. (2017). Understanding the dynamics of Nigeria's urban transition: A refutation of the "stalled urbanisation" hypothesis. *Urban Studies*, pp. 1–7. Available at: http://journals.sagepub.com/doi/abs/10.1177/0042098017712688 [Accessed 6 Feb. 2018].

Gollin, D., Jedwab, R. and Vollrath, D. (2016). Urbanisation with and without Industrialization. *Journal of Economic Growth*, 21(1), pp. 35–70.

Ianchovichina, E. and Lundstrom, S. (2009). What is inclusive growth? *The World Bank*, February 10, 2009. Available at: http://siteresources.worldbank.org/INTDEBTDEPT/Resources/468980-1218567884549/WhatIsInclusiveGrowth20081230.pdf [Accessed 6 Feb. 2018].

Ingelaere, B., Christiaensen, L., De Weerdt, J. and Kanbur, R. (2018). Why secondary towns can be important for poverty reduction—A migrant perspective. *World Development Group, Working Paper 8*. Available at: https://doi.org/10.1016/j.worlddev.2017.12.025 [Accessed 6 Feb. 2018].

Jedwab, R. (2011). Why is African urbanisation different? Evidence from resource exports in Ghana and Ivory Coast. *Job Market Paper (Paris School of Economics and STICERD, London School of Economics)*, pp. 1–60. Available at: http://citeseerx.ist.psu.edu/viewdoc/download?doi=10.1.1.449.7962&rep=rep1&type=pdf [Accessed 6 Feb. 2018].

Kessides, C. (2007). The urban transition in sub-Saharan Africa: Challenges and opportunities. *Environment and Planning C: Government and Policy*, 25(4), pp. 466–485.

Lakner, C. and Milanovic, B. (2015). Global income distribution from the fall of the Berlin Wall to the Great Recession. *Revista de Economía Institucional*, 17(32), pp. 71–128.

McGranahan, G. and Satterthwaite, D. (2014). Urbanisation concepts and trends. *IIED, Working Paper*. London: IIED.

Mitlin, D. and Satterthwaite, D. (2013). *Urban poverty in the global south: Scale and nature*. Oxon: Routledge.

Myers, G. (2011). Why Africa's cities matter. *African Geographical Review*, 30(1), pp. 101–106.

Njoh, A.J. (2003). Urbanisation and development in sub-Saharan Africa. *Cities*, 20(3), pp. 167–174.

Oyeniyi, B.A. (2013). Internal migration in Nigeria: A positive contribution to human development. *ACP Observatory on Migration International Organization for Migration*. Available at: https://publications.iom.int/system/files/pdf/nigeria.pdf [Accessed 6 Feb. 2018].

Parnell, S. and Pieterse, E.A. (2014). *Africa's urban revolution*. London: Zed Books.

Potts, D. (2006). Counter-urbanisation on the Zambian Copperbelt? Interpretations and implications. *Urban Studies*, 42(4), pp. 583–609.

Potts, D. (2008). Recent trends in rural-urban and urban-rural migration in sub-Saharan Africa: The empirical evidence and Implications for understanding urban livelihood insecurity. *Environment, Politics and Development, Working Paper Series 6*. Department of Geography, King's College London.

Potts, D. (2009). The slowing of sub-Saharan Africa's urbanisation: Evidence and implications for urban livelihoods. *Environment and Urbanisation*, 21(1), pp. 253–259.

Potts, D. (2012a). What do we know about urbanisation in sub-Saharan Africa and does it matter? *International Development Planning Review*, 34(1), pp. v–xxii.

Potts, D. (2012b). *Whatever happened to Africa's rapid urbanisation?* [pdf] Africa Research Institute. Available at: http://africaresearchinstitute.org/newsite/wp-content/uploads/2013/03/Whatever-happened-to-Africas-rapid-urbanisation.pdf [Accessed 6 Feb. 2018].

Potts, D. (2013). Rural-urban and urban-rural migration flows as indicators of economic opportunity in sub-Saharan Africa: What do the data tell us? *Migrating out of poverty research programme, Working Paper 9*. Brighton: Migrating out of Poverty Consortium, University of Sussex.

Potts, D. (2016). Debates about African urbanisation, migration and economic growth: What can we learn from Zimbabwe and Zambia? *The Geographical Journal*, 182(3), pp. 251–264.

Potts, D. (2017). Urban data and definitions in sub-Saharan Africa: Mismatches between the pace of urbanisation and employment and livelihood change. *Urban Studies*, pp. 1–22. Available at: https://doi.org/10.1177/0042098017712689 [Accessed 6 Feb. 2018].

Ravallion, M., Chen, S. and Sangraula, P. (2007). New evidence on the urbanisation of global poverty. *Population and Development Review*, 33(4), pp. 667–701.

Sahn, D.E. and Stifel, D.C. (2003). Urban—rural inequality in living standards in Africa. *Journal of African Economies*, 12(4), pp. 564–597.

Satterthwaite, D. (2007). The transition to a predominantly urban world and its underpinnings. *Human Settlements Discussion Paper Series, Theme: Urban change*, 4. London: IIED.

Satterthwaite, D. (2010). Urban myths and the misuse of data that underpin them. *UNU-WIDER, Working Paper 2010/28*. Helsinki: World Institute for Development Economics Research.

Scutaru, L. (2013). Economic development versus sustainable development. *Ecoforum*, 2(1/2), pp. 35–40.

Shifa, M. and Leibbrandt, M. (2017a). Urban poverty and inequality in Kenya. *Urban Forum*, 28(4), pp. 363–385.

Shifa, M. and Leibbrandt, M. (2017b). Profiling multidimensional poverty and inequality in Kenya and Zambia at sub-national levels. *SADLRU, Working Paper Series 209*. Cape Town: SALDRU, UCT. Available at: http://opensaldru.uct.ac.za/bitstream/handle/11090/885/2017_209_Saldruwp.pdf?sequence=3 [Accessed 6 Feb. 2018].

Turok, I. (2013). Securing the resurgence of African cities. *Local Economy*, 28(2), pp. 142–157.

Turok, I. and McGranahan, G. (2013). Urbanisation and economic growth: The arguments and evidence for Africa and Asia. *Environment and Urbanisation*, 25(2), pp. 465–482.

Turok, I. (2017). Informing Africa's urban transformation: A response to Fox *et al.* and Potts. *Urban Studies*, pp. 1–7. Available at: http://journals.sagepub.com/doi/abs/10.1177/0042098017713575 [Accessed 6 Feb. 2018].

United Nations, (2015). *World urbanisation prospects: The 2014 revision* (ST/ESA/SER.A/366). UN Department of Economic and Social Affairs, Population Division.

United Nations General Assembly, (2016). *Draft outcome document of the United Nations Conference on Housing and Sustainable Urban Development (Habitat III)*, 17–20 October, Quito. New York, NY: United Nations.

Weeks, J. (1994). Economic aspects of rural-urban migration. In: J.D. Tarver, ed., *Urbanisation in Africa: A handbook*. Westport, CT: Greenwood Press, pp. 338–407.

World Bank; International Monetary Fund, (2013). *Global monitoring report 2013: Rural-urban dynamics and the millennium development goals*. Washington, DC: The World Bank.

2 Rural bias and urban food security

Jonathan Crush and Liam Riley

Introduction

A review of global food security policy debates in the first decade of the twentieth century showed that the "new international food security agenda" was dominated by a pervasive rural bias that focused almost exclusively on rural hunger and increased support of smallholder agriculture (Crush and Frayne 2011). This pendulum swing was in many ways a reaction to the urban bias that had characterised the global development agenda from the 1970s onwards (Lipton 1977; Jones and Corbridge 2010). Rural bias could also be seen in the food security programmatic statements of international organisations, intergovernmental agencies, regional bodies such as the African Union, and the food security mitigation plans of individual governments. At the same time, despite mounting evidence of rapid urbanisation in the Global South, the "invisible crisis" of food insecurity of urban populations remained a marginal concern at all levels of governance from the global to the local (Crush and Frayne 2010). In part, the invisibility of urban food insecurity was a product of the roadmap laid out in the Millennium Development Goals (MDGs) (Battersby 2017).

The MDGs were ill equipped to address the particular challenges of food insecurity in a rapidly urbanising Africa (Crush and Battersby 2016). Urban poverty reduction was limited to the improvement of slums in the Global South and an arbitrary target of improved conditions for 100 million slum dwellers (Cohen 2014). In reducing the concept of food security to the problem of rural hunger, the MDGs cut short policy debate on urban food and nutrition security and hunger (Fukuda-Parr and Orr 2014; Haddad 2013). Taken separately, the development challenges of urban poverty and food insecurity were inadequately addressed by the MDGs. However, the broader problem was the lack of integrated thinking about these two issues (Battersby 2017). The "siloing" effect of the delineated goals and targets meant that across sectors the big picture of development – including the inter-relatedness of various contributing factors of poverty and underdevelopment – was obscured by perverse incentives to reach specific targets (Fukuda-Parr 2014). The effect was to set back the progress on addressing urban food security concerns that had emerged in the 1990s (Maxwell 1999; Ruel and Garrett 1999; Smith 1998).

This chapter updates the arguments of Crush and Frayne (2011) concerning the pervasive rural bias and anti-urbanism in global, regional, and national responses to the challenge of urban household food insecurity. It examines whether the last decade has brought any significant changes to the dominant discourse and its accompanying sidelining of urbanisation and urban food security in policy debate and formulation. We argue that there are promising cracks in the edifice but that rural bias remains the dominant feature of current thinking about food security policies. Although researchers have begun to press for the urban to be included on the food security agenda, and food to be included on the urban agenda, there has been limited policy uptake to date at the international level and little at all at the municipal governance level.

Locating the urban in the food security agenda

The rural and smallholder agriculture bias which characterised much global thinking about food security in the first decade of the twenty-first century has largely persisted into the second. The Sustainable Development Goals (SDGs) have an enhanced focus on food security (Goal 2) and a new focus on sustainable urbanisation (Goal 11). The objective of Goal 2 is to "end hunger, achieve food security and improved nutrition and promote sustainable agriculture." This brings food security into focus as its own goal, rather than as a subset of the poverty goal as in the MDGs. The first target of Goal 2 is "by 2030, [to] end hunger and ensure access by all people, in particular the poor and people in vulnerable situations, including infants, to safe, nutritious and sufficient food all year round." Even though the target suggests increasing awareness of food security in an urbanising global context, the overall set of targets focus on production and sustainable agriculture. As Battersby notes (2017: 122), "SDG 2 continues to frame the food problem as one of scarcity informed by the knowledge effect of the MDGs." That is to say, the limited view of food security in the MDGs is amplified in the SDGs.

One notable innovation in the SDGs is a stronger emphasis on inter-linkages among development goals and targets, which are connected under the umbrella concept of sustainability. This emphasis has achieved some recognition by the influential Committee on World Food Security (CFS). For example, CFS (2015: 8) quotes Mary Robinson's speech on engaging with the 2030 SDG Agenda in which she noted that "critical to success will be strategies that cut across sectors and adopt people-centred interdisciplinary approaches." While the proposed shift to cross-sectoral approaches to food security holds promise for an *urban* food security agenda, the structure of the goals and targets does little to guarantee such an outcome. While many advocated for a more systemic integration of the goals and targets (Weitz *et al.* 2014), ultimately the MDG structure was repeated with 17 SDGs and 169 targets. As a result, it is likely that "business as usual" will prevail and deeply interconnected issues like urbanisation and food security will continue to be "siloed" in global development discourse and practice (Battersby 2017).

An analysis of the SDGs and related programmatic documents on how to achieve food security reveals a continuing imbalance among the four pillars of food security, with the emphasis still overwhelmingly on food availability. An urban-centred lens on food security suggests that the other major dimensions of food security (access, utilisation, stability, and safety) are critically important in the context of rapid urbanisation. However, food security programming at the global, regional, and national scales continues to be based on a narrow conceptualisation of food security that is poorly equipped to address the growing need for urban solutions. Even when the broad definition of food security is cited, food security is implicitly reduced to production and rural development. Urbanisation is portrayed as threatening food security, cities are reduced to consumer markets, and urbanites are too often assumed to be privileged groups exploiting the work of farmers. In many quarters, urbanisation is even perceived as a threat – leading to more demands on resources, more inequality and conflict, and even the erosion of cultures and morality.

New research perspectives are increasingly challenging the traditional dichotomies of urban–consumer, rural–producer (Lerner and Eakin 2011). As Tacoli and Vorley (2015: 1) note, "our food security narratives are outdated: urban dwellers are not all 'over-consumers'; rural communities are not exclusively producers." It is this lack of complexity – an apparently under-appreciation of geographical nuance – in the debates over the global food security agenda that raises concern about the suitability of current policies for the real challenges facing Africa in the coming decades. The world in 2030 will be much more urbanised than it is in 2017 and many contentious policy decisions will be required to ensure that cities evolve as inclusive, sustainable, and food-secure spaces.

A recent publication entitled "Toward a Food Secure Future" typifies a conservative view of Africa's shifting demographics and leads to policy prescriptions that are inherently anti-urban (Conceicao *et al.* 2016). Using production data that dates back several decades, and pointing to current labour market structures where the majority of African workers are employed in agriculture, the authors advocate policies that focus on small-scale farming to reduce poverty and food insecurity. This policy has the supposed "benefit" of reducing rural to urban migration. The analysis ignores the processes of social change as well as the inevitability of human migration under conditions of social, economic, and environmental change. The rural bias inherent in such advocacy statements continues to determine the priorities of the international food security agenda, as evidenced by the priorities of influential groupings such as the Food and Agriculture Organization (FAO), International Fund for Agricultural Development (IFAD), the World Food Programme (WFP), and the CFS.

Rural bias is particularly evident in the imagery and content of the annual *State of Food Insecurity* (SOFI) and *State of Food and Agriculture* (SOFA) reports of the FAO, WFP, and IFAD. Tellingly, the covers of all three SOFI and four SOFA reports published between 2013 and 2016 feature images of small rural farms and farmers. SOFA 2014 on "Innovation in Family Farming" describes family farms as "stewards of the world's agricultural resources" and claims that they are the source of more than 80% of the world's food supply. Innovation in small farm agricultural practices will "lift farmers out of poverty and help the world achieve food security and sustainable agriculture" (FAO 2014: 93). SOFA 2015 focuses on social protection mechanisms for rural populations, asserting that "poor and food-insecure families depend primarily on agriculture for their livelihoods, and make up a large proportion of the beneficiaries of social protection programmes" (FAO 2015: xvi). The report maintains that "extreme poverty is disproportionately concentrated in rural areas" and that 52% of the rural population in sub-Saharan Africa (SSA) lives in extreme poverty and food insecurity (FAO 2015: xii). The 29% of urban dwellers identified by the report as living in extreme poverty are systematically ignored (FAO 2015: 9). SOFA 2016 addresses the impact of climate change on food security, again with a particular emphasis on production by smallholder farmers. The complex effects of climate change on urban food security are not addressed (Frayne *et al.* 2012). SOFA 2017 purports to focus on rural–urban linkages but its main policy objective is "leveraging food systems for an inclusive rural transformation that leads to prosperity and the eradication of hunger and poverty" (FAO 2017a: xii). In other words, while the realities of rapid urbanisation are increasingly acknowledged, the process is still largely seen in terms of its implications for rural poverty and food insecurity, and the need to support smallholder production.

All three SOFI reports display a broader appreciation of the complexity of food security but are ultimately unable to escape the straitjacket of rural bias. SOFI 2013 provides a useful summary of different metrics for the analysis of all four dimensions of food security but focuses only on regional and national variations in food security, neglecting finer intra-national scales of analysis (FAO/IFAD/WFP 2013). SOFI 2014 on "Strengthening the Enabling Environment for Food Security and Nutrition" calls for food security and nutrition to be "at the top of the political agenda and creating an enabling environment for improving food security and nutrition through adequate investments, better policies, legal frameworks, stakeholder participation and a strong evidence base" (FAO/IFAD/WFP 2014: 55). The call for national food security policies is certainly welcome, though the concrete suggestions are again compromised by rural bias. For example, SOFI 2014 suggests a policy environment which combines "immediate hunger relief interventions with long-term actions for sustainable growth, especially in agriculture and the rural economy" (FAO/IFAD/WFP 2014: 18). SOFI 2015 charts regional shifts and variations in hunger and devotes a whole section to "the contribution of family farming and smallholder agriculture to food security and nutrition. The report asserts that "to accelerate progress in improving access to food by the poor, lagging regions, particularly SSA, will increasingly have to transform their agricultural policies to significantly improve agricultural productivity and increase the quantity of food supplied by family farmers" (FAO/IFAD/WFP 2015: 33).

The outputs of the CFS are also representative of the policy priorities of the current international food security agenda. The *Global Strategic Framework for Food Security and Nutrition*, for example, provides much evidence of rural bias (CFS 2016a). "Urbanisation and rural-urban migration" are asserted to be a "cause" of hunger and malnutrition (CFS 2016a: 9) without any further explanation or evidence of this supposed causal relationship. The objective of the strategic framework may be to address the food security concerns of the rural poor, but recent migrants and long-time urban residents are missing from the picture or presumed to be food secure. A later section of the *Global Framework* on "issues requiring further attention" suggests that urbanisation can be countered by "boost[ing] rural development to strengthen food security in the context of rural-urban migration" (CFS 2016a: 67). The implication is that production is threatened by the loss of rural agricultural labour through migration from rural to urban areas.

The *Global Framework* falls short of addressing the role of cities and other sub-national entities in ensuring food security and nutrition. The framework is based on the "Five Rome Principles for Sustainable Global Food Security" that "provide a powerful strategic underpinning for coordinated action by all stakeholders at global, regional, and country levels" (CFS 2016a: 12). The "national" level is equated to the local scale, even in large countries where megacities include millions of residents. The *Global Framework* mentions the sub-national scale only under recommendations at country level to "coordinate strategies and actions with local levels of government" (CFS 2016a: 54). The conflation of "national" with "local" is a very different orientation than in the urban agenda where the role of sub-national governments is well recognised. Overall, the international food security agenda continues to do a poor job of conceptualising cities and rural–urban dynamics.

The focus on agriculture in food security policy discourses has remained even more pronounced in Africa due to the enduring perception of African societies as predominantly rural and the relatively high percentage of employment and Gross Domestic Product (GDP) in the agricultural sectors of African countries. The FAO's (2017b) optimistic prognosis on positive trends in food security and agriculture calls on governments to maintain the momentum through greater support for smallholder agriculture. The conflation of food security with agricultural production is evident throughout the report. The African Union's (AU) (2017) Food Security Priority also re-affirms rural smallholder production as key to ending food insecurity in Africa:

> The continent can extricate itself from the vicious cycles of drought, hunger and famine by putting emphasis on the right policies to improve productivity of smallholders, more effective nutrition policies, targeting especially children, building households' ability to cope with shocks, empowering women, the youth and persons with disabilities and accelerating rural infrastructure and value addition.
>
> (AU 2017: no page)

The Comprehensive Africa Agriculture Development Programme (CAADP) was an initiative by the AU and New Partnership for Africa's Development (NEPAD) and formed the core of the regional governance agenda for food security since 2004. CAADP apparently "reflects African governments' recognition of agriculture as central for the alleviation of poverty and hunger" (UN 2017a). In 2014, the Malabo Declaration marked the tenth anniversary of the CAADP and involved a pledge by African heads of state and government to "end hunger by 2025 by at least doubling current agricultural productivity levels, reducing postharvest losses and waste by at least half the current level, and reducing stunting to 10% and underweight to 5%" (FAO 2017c: 11).

The Malabo Declaration is aligned with the AU's Agenda 2063, which sets out a vision for *The Africa We Want* (AU 2015). In the first ten-year implementation plan, Aspiration 1 includes "a prosperous Africa based on inclusive growth and sustainable development." There are several targets related to ending hunger including an 80% decrease in the incidence of hunger and malnutrition; a 50% reduction in "all forms of malnutrition, maternal, child and neonatal mortalities"; and GDP growth that will "provide the resources and the medium for eliminating poverty and hunger" (AU 2015: 43–44). The strategies for achieving these goals include a mix of agricultural and macro-economic strategies. Agricultural strategies include the implementation of CAADP, strategies that will boost the productivity of rural households, and the promotion of high-nutrition and drought-resistant crops. These are blended with strategies aimed at making food more accessible and affordable (through strengthening markets, food trade, and strategic food reserves), and more nutritious. These policies have the potential to benefit people in both urban and rural areas. However, the rural bias and isolation of food security challenges from urbanisation are evident in the Declaration.

Locating food in the urban agenda

The obstacles to creating policies that can address the challenge of urban food security are not limited to anti-urban biases of the food security agenda. There is a complementary absence of food security in discourses and development interventions in the urban agenda. The new urban SDG – Goal 11 – promisingly aims to "make cities and human settlements inclusive, safe, resilient and sustainable" (UN 2017b). However, food is altogether absent from the urban SDG, which includes ten targets related to housing, transportation, participatory planning, disaster risk reduction, and other issues that may be related to food but do not specifically serve the food security agenda in cities. The effect of defining a set of urban issues of concern is to define other issues, such as food security, as *not* inherently urban. Battersby (2017: 124) notes that "ironically, having a specifically urban goal may have led to a lack of engagement [with SDG 2]." And yet, by 2030 the global population will be even more urbanised and the need for a global food agenda that recognises the needs of poor urban consumers will be even more urgent (Crush 2016).

The global picture of urbanisation presented by UNHABITAT appears to be as unengaged with food security as the international food security agenda is with the urban. UNHABITAT has traditionally avoided inclusion of urban food issues in its programming priorities. Most recently, the 2015 African Urban Agenda document prepared for HABITAT III discussions omits any reference to food security (UNHABITAT 2015). The continuing omission of food from UNHABITAT's brief is indicative of the separation of food security from the urban agenda at the global and continental levels. The official African regional declaration for HABITAT III did not explicitly name food as an urban challenge or development priority, even though issues like housing and water were mentioned (UN 2016). The declaration is explicitly "guided by the African Union's Agenda 2063" (UN 2016: 2). The urban vision in Agenda 2063 does not make reference to food or food systems (AU 2015).

The 2016 *World Cities Report* potentially signifies a new UNHABITAT sensitivity when it notes that a "shift towards an increasingly urbanised world constitutes a force which can be harnessed for a more sustainable development trajectory. This dramatic shift towards urban life has profound implications for energy consumption, politics, *food security* [emphasis added] and human progress" (Moreno *et al.* 2016: 29). Yet, the report then goes on to mention food and food security primarily in terms of food production, for example in the effect of urban sprawl on the loss of farmland (Moreno *et al.* 2016: 51, 71, 129), the effects of climate change on agriculture and hence the urban food supply (Moreno *et al.* 2016: 181), and the potential for global food security to benefit from biodiversity in cities (Moreno *et al.* 2016: 107). The conventional framing of food security as a non-urban issue is evident in the statement that "even *seemingly unrelated issues such as food security* [emphasis added] and rural water supplies are closely tied to the economic growth and prosperity of cities" (Moreno *et al.* 2016: 152).

The New Urban Agenda (NUA), accepted at the HABITAT III conference in 2016, is the focal point of the global urban agenda until 2036 (UNHABITAT 2017). HABITAT III included many municipal and civil society representatives and, as a result, the NUA places more emphasis on sub-national actors than the SDGs (Parnell 2016). It also strikes an optimistic tone about the potential for sustainable cities to optimise the benefits of new technologies and models of inclusive governance for the conservation of natural resources, the preservation of eco-systems, and the promotion of equitable growth. Although there was some resistance to including food security in the NUA, sustained lobbying by various non-governmental agencies saw it named in 12 of the 175 articles of the document. In most cases, food or food and nutrition security are simply included in lists of desirable public goods, services, and outcomes. One section stands out for advocating a broader focus on urban food security (UNHABITAT 2017: 39):

> We will promote the integration of food security and the nutritional needs of urban residents, particularly the urban poor, in urban and territorial

> planning, in order to end hunger and malnutrition. We will promote coordination of sustainable food security and agriculture policies across urban, peri-urban and rural areas to facilitate the production, storage, transport and marketing of food to consumers in adequate and affordable ways in order to reduce food losses and prevent and reuse food waste. We will further promote the coordination of food policies with energy, water, health, transport and waste policies, maintain the genetic diversity of seeds, reduce the use of hazardous chemicals and implement other policies in urban areas to maximize efficiencies and minimize waste.

While a focus on promoting the "integration" and "needs of urban residents" is promising, the policy solution primarily ties food security to production and the reduction of food waste rather than to the full spectrum of actions that would promote food-secure cities.

A range of non-governmental organisations and research organisations have continued to advocate productionist solutions to household food insecurity in the form of urban agriculture (Lee-Smith 2010; Redwood 2009). Some national and municipal governments have incorporated urban agriculture as the focus of their food and nutrition security planning. Battersby *et al.* (2015: 2) argue that in South Africa, for example, "the promotion of urban agriculture has been the major food security intervention at the urban scale. It has consistently been national government's lens for engaging the urban food security challenge." However, as a response to urban food insecurity, urban agriculture "does not provide an adequate response to the urban challenge." Expecting the urban poor, who have the least access to the resources (money, land, tools, seed, knowledge, equipment) necessary to establish successful agricultural ventures, to "grow their own" in order to lift themselves out of poverty, "fails to recognise the massive barriers constraining urban agriculture in South African cities" (Battersby *et al.* 2015: 2). Similar arguments have been made for East Africa where urban agriculture has been seen as the primary policy response to food insecurity (Brown 2015). The evidence-based critique of urban agriculture has been replicated in other southern African countries and in other parts of Africa and the Global South (Badami and Ramankutty 2015; Frayne *et al.* 2016; Warren *et al.* 2015; Zezza and Tasciotti 2010). Urban agriculture can be beneficial, but there is little evidence that it is effective in targeting the needs of the most vulnerable urban residents.

While the research on urban food security in Africa continues to accumulate through the work of the African Food Security Urban Network (AFSUN), Hungry Cities Partnership (HCP), the Consuming Urban Poverty (CUP) Project and others, there is little evidence that this body of work is shifting the policy priorities of the African regional and national food security and urban development agendas. However, there are signs that food could become more prominent within an African urban and food security agenda over the next decade through a growing policy emphasis on nutrition issues. Nutrition highlights the food security pillar of utilisation and nutritionists have long argued that the MDG focus on undernutrition and hunger understated the importance of other kinds of food insecurity, such as micronutrient deficiencies and the burden of overnutrition (Haddad 2013). Nutrition has begun to make an increasingly important contribution to the global food agenda (Haddad *et al.* 2015; Global Panel 2016; FAO/IFAD/WFP 2015). The integration of the African Regional Nutrition Strategy in the AU Agenda 2063 is further evidence of the increasing prominence of nutrition (AU 2017). These developments reinforce the argument that food security is a multi-sectoral problem that is far more complicated than simply growing more food. On the other hand, much of the nutritionist agenda does still tend to be production-focused, for example in advocating for nutrition-sensitive agriculture, with less attention paid to nutrition needs in cities (Jaenicke and Virchow 2013).

Nutrition narratives are increasingly linked to urbanisation through concerns about the double burden of nutrition with rising obesity rates and consequential rises in non-communicable disease such as diabetes and heart disease. The High Level Panel of Experts for the CFS starkly stated in the first "critical and emerging issues" paper that "urbanisation leads to obesogenic diets and behaviours" because of household income growth and the increased consumption of animal sourced food (CFS 2017: 7). Such narratives risk oversimplification of the causal connections between urbanisation and changing diets because the urban food systems are not only shaped by consumer demand but also by the food systems that supply certain foods to the cities and other aspects of urban culture that shape certain preferences (Crush and Battersby 2016; Tacoli and Vorley 2015; Bloem and de Pee 2017). Obesity in urban areas is found among low- and high-income communities (Battersby 2017). The emerging health crisis of malnutrition is embedded across environmental, economic, and cultural dynamics and cannot be addressed using a non-dynamic concept of urbanisation.

Other examples of an emerging policy awareness of the multi-faceted challenges of urbanisation for food and nutrition security can be found in recent contributions from the World Bank, the International Food Policy Research Institute (IFPRI), and the Milan Urban Food Policy Pact. The World Bank's 2017 report on African cities notes the effects of high food prices in cities, where "city dwellers pay around 35 percent more for food in Africa than in low-income and middle-income countries elsewhere: a premium that looms larger given the high share of African household incomes that goes to food" (Lall *et al.* 2017: 23). The report is primarily concerned with the negative impact on economic development of high urban food costs, but it is an important contribution to regional debates that tend to overlook the ramifications of a broken food system. The World Bank's decade-long obsession with smallholder agriculture may also shift in future as its Food and Agriculture Global Practice division is currently (mid-2017) preparing a scoping report on *Food Systems for an Urbanising World* as the first step in deciding how the Bank can "advance a transformative agenda in support of urban food systems" (The World Bank 2017: ix).

After years of focusing on rural food insecurity, IFPRI's latest Global Food Policy Report for 2017 notes that "rapid urbanisation, particularly in developing countries, is a critical ongoing trend shaping food security and nutrition that will continue in 2017 and beyond" (IFPRI 2017: 9). The report itself explores the current state of knowledge on a range of issues relating to food security in the cities and could serve as a flagship for a new policy agenda at the international, regional, national, and municipal levels. The CFS 43rd Session in 2016 was again overwhelmingly concerned with rural transformation but did table a document for discussion entitled "Urbanisation, Rural Transformation and Implications for Food Security and Nutrition: Key Areas for Policy Attention and Possible Roles for CFS" (the product of an online forum hosted by the FAO) (CFS 2016b: 13). The CFS has also created an "Open Ended Working Group on Urbanisation and Rural Transformation" "to agree on a process to compile experiences and effective policy approaches for addressing food security and nutrition in the context of changing rural-urban dynamics" (CFS 2016b: 13). What this process might actually look like is still to be seen. Finally, the Milan Urban Food Policy Pact is attempting to place urban food issues on the municipal policy agenda. The Pact began with European cities and as of February 2018 includes international representation and 25 African cities out of 163 signatory cities (MUFPP undated).

Conclusion

In contrast to some highly urbanised countries and major cities in Latin America, there is little awareness of the importance of crafting policy responses for managing and mitigating the growing crisis of urban food insecurity in Africa (Haysom 2015). Those countries which have formulated national food and nutrition security plans (such as Kenya, South Africa, and Uganda) fail to recognise the magnitude of the challenge, and as result they tend to get caught up in viewing food security as an agricultural and rural development challenge. At the city level, there are few, if any, examples of coherent policy responses to the management of urban food systems in the interests of the poor and food insecure. Cities like Cape Town, South Africa, that have initiated a process to develop a food security strategy have later abandoned the effort (Battersby *et al.* 2014; Haysom *et al.* 2017). A possible alternative route would see the mainstreaming of food security into national and local urban development planning and governance. However, despite the ubiquity of food in African cities, food security is notable for its absence from local and national urban development agendas (Brown 2015; Haysom 2015; Smit 2016).

In this chapter, we have examined the global and African regional context within which both urban development and food security agendas in Africa are framed. We argue that the pervasive rural bias and anti-urbanism identified in the international and regional food security agendas in the first decade of the twenty-first century have persisted into the second. The SDGs, like the MDGs before them, provide few grounds for optimism going forward, nor do the priorities of UN agencies such as the FAO, IFAD, and the WFP or the CFS. If urban food security is addressed in a substantive manner, it will probably only be indirectly through the actions of the influential global nutrition lobby. Despite the promise of the NUA, there are grounds for caution about its ability to seriously and systematically formulate and promote a coherent set of policy interventions that reach much beyond the tired mantra of urban agriculture. That said, continued research, advocacy efforts (such as the recent Bellagio Declaration), and initiatives to put food onto urban policy agendas at the local level should continue in order to lay the foundations for innovative and rights-regarding policy responses for the time when Africa's urban marginalised and food insecure force themselves onto the governance agenda.

Acknowledgements

We wish to acknowledge the support of the SSHRC and IDRC for the Hungry Cities Partnership, the ESRC-DFID for its support of the Consuming Urban Poverty (CUP) project, and the SSHRC for its support of the CUP2 Project.

References

AU, (2015). *Agenda 2063: The Africa we want: Strategic framework for inclusive growth and sustainable development: Ten-year implementation plan 2014–2023.* Addis Ababa: African Union Commission.

AU, (2017). *AUC priorities: Food security.* African Union Commission. Available at: www.au.int/web/en/auc/priorities/food-security [Accessed 28 Feb. 2018].

Badami, M. and Ramankutty, N. (2015). Urban agriculture and food security: A critique based on an assessment of urban land constraints. *Global Food Security*, 4, pp. 8–15.

Battersby, J. (2017). MDGs to SDGs—new goals, same gaps: The continued absence of urban food security in the post-2015 Global Development Agenda. *African Geographical Review*, 36, pp. 115–129.

Battersby, J., Haysom, G., Kroll, F. and Tawodzera, G. (2015). *Looking beyond urban agriculture: Extending urban food policy responses. Sustainable Cities Policy Brief.* Johannesburg: South African Cities Network.

Battersby, J., Haysom, G., Tawodzera, G., McLachlan, M. and Crush, J. (2014). *Food system and food security study.* Cape Town: City of Cape Town.

Bloem, S. and de Pee, S. (2017). Developing approaches to achieve adequate nutrition among urban populations requires an understanding of urban development. *Global Food Security*, 12, pp. 80–88.

Brown, A. (2015). Sustaining African cities: Urban hunger and sustainable development in East Africa. *International Journal of Environmental, Cultural, Economic, and Social Sustainability: Annual Review*, 11, pp. 1–12.

Cohen, M. (2014). The city is missing in the millennium development goals. *Journal of Human Development and Capabilities*, 15, pp. 261–274.

CFS, (2015). *Report on forty second session: Making a difference in food security and nutrition.* Rome: Committee on World Food Security. Available at: www.fao.org/3/a-mo960e.pdf [Accessed 23 Feb. 2018].

CFS, (2016a). *Global strategic framework for food security & nutrition. Fifth Version*. Rome: Committee on World Food Security. Available at: www.fao.org/3/a-mr173e.pdf [Accessed 23 Feb. 2018].

CFS, (2016b). *Forty-third session: Making a difference in food security and nutrition*. Rome: Committee on World Food Security. Available at: www.fao.org/3/a-ms023e.pdf [Accessed 23 Feb. 2018].

CFS, (2017). *Second note on critical and emerging issues for food security and nutrition*. Rome: Committee on World Food Security. Available at: www.fao.org/fileadmin/user_upload/hlpe/hlpe_documents/Critical-Emerging-Issues-2016/HLPE_Note-to-CFS_Critical-and-Emerging-Issues-2nd-Edition_27-April-2017_.pdf [Accessed 23 Feb. 2018].

Conceicao, P., Levine, S., Lipton, M. and Warren-Rodriquez, A. (2016). Toward a food secure future: Ensuring food security for sustainable human development in sub-Saharan Africa. *Food Policy*, 60, pp. 1–9.

Crush, J. (2016). *Hungry cities in the global south. Hungry Cities Discussion Paper No. 1*. Waterloo and Cape Town: Hungry Cities Partnership.

Crush, J. and Battersby, J., eds. (2016). *Rapid urbanisation, urban food deserts and food security in Africa*. Cham: Springer.

Crush, J. and Frayne, B. (2011). Urban food insecurity and the new international food security agenda. *Development Southern Africa*, 28, pp. 527–544.

Crush, J. and Frayne, B. (2010). *The invisible crisis: Urban food security in Southern Africa. Urban Food Security Series No. 1*. Cape Town: African Food Security Urban Network.

FAO, (2014). *The state of food and agriculture: Innovation in family farming*. Rome: UN.

FAO, (2015). *The state of food and agriculture: Social protection and agriculture: Breaking the cycle of rural poverty*. Rome: Food and Agriculture Organization of the UN.

FAO, (2017a). *The state of food and agriculture: Leveraging food systems for rural transformation*. Rome: Food and Agriculture Organization of the UN.

FAO, (2017b). *The future of food and agriculture: Trends and challenges*. Rome: Food and Agriculture Organization of the UN.

FAO, (2017c). *Regional overview of food security and nutrition in Africa 2016: The challenges of building resilience to shocks and stresses*. Accra: UN.

FAO/IFAD/WFP, (2013). *The state of food insecurity in the world: The multiple dimensions of food security*. Rome: United Nations.

FAO/IFAD/WFP, (2014). *The state of food insecurity in the world: Strengthening the enabling environment for food security and nutrition*. Rome: United Nations.

FAO/IFAD/WFP, (2015). *The state of food insecurity in the world: Meeting the 2015 international hunger targets: Taking stock of uneven progress*. Rome: United Nations.

Frayne, B., McCordic, C. and Shilomboleni, H. (2016). The mythology of urban agriculture. In: J. Crush and J. Battersby, eds., *Rapid urbanisation, urban food deserts and food security in Africa*. Cham: Springer, pp. 19–32.

Frayne, B., Moser, C. and Ziervogel, G. (2012). *Climate change, assets and food security in Southern African cities*. London: Earthscan.

Fukuda-Parr, S. (2014). Global goals as a policy tool: Intended and unintended consequences. *Journal of Human Development and Capabilities*, 15, pp. 118–131.

Fukuda-Parr, S. and Orr, A. (2014). The MDG hunger target and the competing frameworks of food security. *Journal of Human Development and Capabilities*, 15, pp. 147–160.

Global Panel, (2016). *Food systems and diets: Facing the challenges of the 21st century*. London: Global Panel on Agriculture and Food Systems for Nutrition.

Haddad, L. (2013). How should nutrition be positioned in the post-2015 agenda? *Food Policy*, 43, pp. 341–352.

Haddad, L., Achadi, E., Bendech, M.A., Ahuju, A., Bhatia, K., Bhutta, K., Blossner, M., Borghi, E., Colecraft, E., Eriksen, M.d'O.K., Fanzo, J., Flores-Ayala, R., Fracassi, P., Kimani-Murage, E., Koukoubou, E.N., Krasevec, J., Newby, H., Nugent, R., Oenema, S., Martin-Preval, Y., Randel, J., Requejo, J., Shyam, T., Udomkesmalee, E. and Reddy, K.S. (2015). The global nutrition report 2014: actions and accountability to accelerate the world's progress on nutrition. *Journal of Nutrition*, 145, pp. 663–671.

Haysom, G. (2015). Food and the city: Urban scale food system governance. *Urban Forum*, 26, pp. 263–281.

Haysom, G., Crush, J. and Caesar, M. (2017). *The urban food system of Cape Town, South Africa. Hungry Cities Report No. 3.* Waterloo and Cape Town: Hungry Cities Partnership.

IFPRI, (2017). *Global food policy report 2017.* Washington, DC: International Food Policy Research Institute.

Jaenicke, H. and Virchow, V. (2013). Entry points into a nutrition-sensitive agriculture. *Food Security*, 5, pp. 679–692.

Jones, G. and Corbridge, S. (2010). The continuing debate about urban bias: The thesis, its critics and its implications for poverty-reduction strategies. *Progress in Development Studies*, 10, pp. 1–18.

Lall, S. Henderson, J. and Venables, A. (2017). *Africa's cities: Opening doors to the world.* Washington, DC: The World Bank.

Lee-Smith, D. (2010). Cities feeding people: An update on urban agriculture in equatorial Africa. *Environment & Urbanisation*, 22, pp. 483–499.

Lerner, A. and Eakin, H. (2011). An obsolete dichotomy? Rethinking the rural-urban interface in terms of food security and production in the global south. *Geographical Journal*, 177, pp. 311–320.

Lipton, M. (1977). *Why poor people stay poor: Urban bias in world development.* Canberra: ANU Press.

Maxwell, D. (1999). The political economy of urban food security in sub-Saharan Africa. *World Development*, 27, pp. 1939–1953.

MUFPP, (2017). *Milan urban food policy pact.* Available at: www.milanurbanfoodpolicypact.org [Accessed 23 Feb. 2018].

Moreno, E., Arimah, B., Otieno Otieno, R., Mbeche-Smith, U., Klen-Amin, A. and Kamiya, M. (2016). *Urbanisation and development: emerging futures: World cities report 2016.* Nairobi: UNHABITAT.

Parnell, S. (2016). Defining a global urban development agenda. *World Development*, 78, pp. 529–540.

Redwood, M., ed., (2009). *Agriculture in urban planning: Generating livelihoods and food security.* Ottawa: IDRC.

Ruel, M. and Garrett, J. (1999). Overview of special Issue: Urban challenges to food and nutrition security. *World Development*, 27, pp. 1885–1889.

Smit, W. (2016). Urban governance and urban food systems in Africa: Examining the linkages. *Cities*, 58, pp. 80–86.

Smith, D. (1998). Urban food systems and the poor in developing countries. *Transactions of the Institute of British Geographers*, 23, pp. 207–219.

Tacoli, C. and Vorley, B. (2015). *Reframing the debate on urbanisation, rural transformation and food security. IIED Briefing.* Available at: http://pubs.iied.org/17281IIED [Accessed 28 Feb. 2018].

UN, (2016). *Abuja declaration for the United Nations Conference on housing and sustainable urban development (Habitat III): Africa's priorities for the New Urban Agenda.* Abuja: United Nations.

UN, (2017a). *Comprehensive Africa agriculture development programme (CAADP).* Office of the Special Advisor to Africa. Available at: www.un.org/en/africa/osaa/peace/caadp.shtml [Accessed 23 Feb. 2018].

UN, (2017b). *Transforming our world: The 2030 agenda for sustainable development.* Available at: https://sustainabledevelopment.un.org/post2015/transformingourworld [Accessed 23 Feb. 2018].

UNHABITAT, (2015). *Towards an African urban agenda.* Nairobi: UNHABITAT and UNECA.

UNHABITAT, (2017). *New Urban Agenda.* Quito: UNHABITAT Secretariat.

Warren, E., Hawkesworth, S. and Knai, C. (2015). Investigating the association between urban agriculture and food security, dietary diversity, and nutritional status: A systematic literature review. *Food Policy*, 53, pp. 54–66.

Weitz, N., Nilsson, M. and Davis, M. (2014). A nexus approach to the post-2015 agenda: Formulating integrated water, energy, and food SDGs. *SAIS Review*, 34, pp. 37–50.

The World Bank. (2017). *Food systems for an urbanising world. Knowledge product.* Washington, DC: The World Bank.

Zezza, A. and Tasciotti, L. (2010). Urban agriculture, poverty, and food security: Empirical evidence from a sample of developing countries. *Food Policy*, 35, pp. 265–273.

3 Linking urban food security, urban food systems, poverty, and urbanisation

Jane Battersby and Gareth Haysom

Introduction

The concept of food security has evolved from its first articulation as a "secure, adequate and suitable supply of food for everyone" (Weingartner 2004: 4) expressed at the 1943 Hot Springs conference. Most notably, it shifted to "the availability at all times of adequate world food supplies of basic food-stuff . . . to sustain a steady expansion . . . and to offset fluctuations in production and prices" (Maxwell 1996: 169) in 1976, and to the commonly used 1996 FAO definition of food security as "the situation that exists when all people, at all times, have physical and economic access to sufficient safe and nutritious food that meets their dietary needs and food preferences for an active and healthy life" (FAO 1996).

As a result of this refining of the concept of food security, the vast majority of food security and food poverty work in the African urban context now focuses at the household scale. While this is a useful shift in scale of analysis from the global and national scale of earlier food security work, this chapter argues that the adoption of a predominantly household scale of analysis and action has obscured the systemic drivers of urban food insecurity and has depoliticized food security.

Although the conceptual framing of food security has gradually shifted from a preoccupation with food availability to the inclusion of multiple dimensions of food security, food security policy and programmatic responses have lagged behind. As Lang and Barling note (2012: 313) food security responses have

> suffered from more than just the common policy ailment of a mismatch between evidence and policy. It is dominated by a discourse emanating from an analysis first charted scientifically in the early to mid-20th century. This is that food insecurity must be centrally addressed by producing more food.

This production orientation has created a rural bias in food security programming and policy.

At the urban scale three distinct foci are evident. First, a production-driven mind-set persists, resulting in project-based interventions such as urban agriculture. Second, interventions and policies focus on the individual or household, resulting in efforts to improve livelihoods. Here interventions such as social protection and income generation initiatives are seen as viable and useful policy approaches. Third, the production and livelihood interventions have a distinct project focus and are targeted at the individual or household level. The result is that the broader food system failings and challenges associated with the urban food system, and how these intersect with the multiple other modes of urban functioning, are either overlooked or disregarded. Kirsten (2012) has termed these interventions as "second class" interventions that seek to mitigate the negative impacts of the prevailing food system, rather than directly engaging the largely private sector-driven structural problems of the food system.

This chapter argues that although the 1996 FAO definition of food security has been adopted by national governments and global agencies, the limited interpretation of the definition and its associated "four pillars" of food security has meant that the drivers of urban food insecurity and its connections to the multi-dimensional nature of urban poverty are poorly understood. The chapter is therefore structured around the FAO's four pillars of food security and argues that if these are viewed through an urban lens, it is possible to reconceptualize urban food security and consider a wider range of policy, planning, and programmatic interventions.

Refocusing the four pillars in the urban African context

According to the FAO, the achievement of food security is based on four pillars, or critical dimensions: availability, access, utilization, and stability. The FAO define these as follows:

> ***Food availability***: The availability of sufficient quantities of food of appropriate quality, supplied through domestic production or imports (including food aid).
>
> ***Food access***: Access by individuals to adequate resources (entitlements) for acquiring appropriate foods for a nutritious diet. Entitlements are defined as the set of all commodity bundles over which a person can establish command given the legal, political, economic and social arrangements of the community in which they live (including traditional rights such as access to common resources).
>
> ***Utilization***: Utilization of food through adequate diet, clean water, sanitation and health care to reach a state of nutritional well-being where all physiological needs are met. This brings out the importance of non-food inputs in food security.
>
> ***Stability***: To be food secure, a population, household or individual must have access to adequate food at all times. They should not risk losing access to food as a consequence of sudden shocks (e.g. an economic or climatic crisis) or cyclical events (e.g. seasonal food insecurity). The concept of stability can therefore refer to both the availability and access dimensions of food security.
>
> (FAO 2006)

Working with these four pillars, it is clear that although food insecurity primarily manifests at the household scale, its causes extend well beyond this scale. Therefore, it should be apparent that the policies, planning, and programming produced to address food security should extend their scale of intervention beyond the household scale. The trap has been that the primary scale of analysis has become the primary scale of intervention.

Extending the scale of intervention beyond the household scale requires a far more critical engagement with the food system. This chapter uses the HLPE (High Level Panel of Experts on Food Security and Nutrition) definition of a food system: "A food system gathers all the elements (environment, people, inputs, processes, infrastructures, institutions, etc.) and activities that relate to the production, processing, distribution, preparation and consumption of food, and the outputs of these activities, including socioeconomic and environmental outcomes" (HLPE 2014: 12). Food system activities are legitimized, enabled, or regulated through policies, institutions, stakeholder actions, practice, governance approaches, and external pressures. The principal challenge when seeking to address food security at the urban scale is then how the food system and the four pillars of food security intersect, enable, or constrain the urban food system and the outcomes that this process generates.

The following sub-sections consider the ways in which an expanded understanding of the four pillars in the African urban context can provide an entry point for novel interventions to alleviate food insecurity.

Availability

In the urban food security context, improved availability has often been seen as a corrective to limited accessibility. In other words, if people cannot access food (usually by monetary means) they should be encouraged and enabled to grow their own food. In their systematic review of urban agriculture's contribution to food and nutrition security in multiple countries, Warren *et al.* (2015) questioned the universal nature of claims promoting urban agriculture's contribution to food security. Globally (Badami and Ramankutty 2015) and in South Africa (SACN 2015), evidence suggests that it is not the poor who are in fact the urban growers. These findings call for far more nuance in how urban agriculture is promoted and more importantly, questions the assertion that "production as the solution" will address food security.

The focus on urban agriculture needs to be understood in the context of wider agricultural policy to ensure food security through ensuring availability of staple crops. There is a growing critique of the single staple crop focus that has epitomized the production focus of earlier food security approaches in a number of countries and sectors. In Zambia, for example, global organizations such as Scaling Up Nutrition (SUN) and HIVOS[1] have challenged the primacy of maize as a food security determinant (Harris *et al.* 2017). The argument is that this provides reduced urban hunger but does not ensure food and nutrition security. This argument is substantiated by Chapter 17.

It is increasingly evident that urban areas are experiencing the triple burden of malnutrition, with undernutrition, micronutrient deficiencies, and overweight, obesity, and associated diet-related non-communicable diseases coexisting in households (Gómez *et al.* 2013). This was acknowledged by the World Health Organization (WHO) and FAO as early as 2002: "Given the rapidity with which traditional diets and lifestyles are changing in many developing countries, it is not surprising that food insecurity and undernutrition persist in the same countries where chronic diseases are emerging as a major epidemic" (WHO/FAO 2002: 8). The 2015 Global Nutrition Report flags high levels of adult obesity in Africa, with 33% of the African adult population being overweight, and a further 11% obese. In southern Africa the corresponding figures are 52% and 26% respectively (Global Nutrition Report 2015a; Global Nutrition Report 2015b). It is not sensible to attribute this shift simply to an agricultural sector orientation to ensure the available of a single staple crop; however, it is clear that a focus on availability alone is not achieving food and nutrition security.

Within Sustainable Development Goal 2 (the "Food SDG") the framing of food security includes addressing both hunger and malnutrition in all its forms (United Nations 2015). This represents a shift in global discourse to focus not just on the basic caloric or protein sufficiency of food made available by the food system, but also the quality of diet made available (Rose *et al.* 2013).

Given the nature of dietary changes, this chapter suggests that the concept of availability needs to be extended to raise questions about the relative balance of types of food being made available within the food system, and why certain types of food are more available than others. Framed this way, the issue of availability raises some fundamental questions about the drivers of the food system.

Accessibility

One of the critical moments in the evolution of the concept of food security was the work of Sen (1981) and his "entitlement thesis," which shifted emphasis from natural causes of food insecurity (drought, pests, etc., within the productive realm) to social, economic, and political causes of vulnerability. Such reconceptualization shifted attention to individual-specific hunger, a view that served to reinforce food security strategies based on poverty reduction, food price, and social protection policies (Barrett 2010: 825).

The concept of accessibility as articulated in the FAO's documents is clearly informed by Sen's work on entitlements, which focuses on the household's capacity to access food. Sen's framework articulates a wide "entitlement set" that individuals and households are able to draw on in order to meet their needs. These are categorized into four types of entitlement: production-based (growing food); trade-based (buying food); own labour-based (working for food); and inheritance- or transfer-based (being given food). While the "Accessibility Pillar" clearly draws on these ideas, it is generally simplified within FAO documents to indicate only physical and economic access to food. In large-scale national food security assessments this is commonly simplified still further to whether or not a household has sufficient income.

Both economic and physical access, and the relationship between these two, are essential. It is, however, not sufficient to look at whether a household has enough money to buy food. Place (in terms of the household and the immediate neighbourhood) and scale (in terms of context, policy focus, and governance) are particularly important when considering food access and the other pillars of food security (Battersby *et al.* 2017). Understanding the location of the sources of affordable and nutritious food is also important, specifically in relation to the lived geographies of urban residents. This includes the geographies of home, work, and travel.

It is also important to understand the pricing and food sector business strategies and the composition of the food retail environment. Wiggins and Keats's (2015) analysis of food pricing in Brazil, China, Mexico, and Korea found that fruit and vegetable prices had risen by between 55% and 91% between 1990 and 2012, but that the prices of four of the six processed foods tracked had shown price falls since 1990. It has been argued that this is the outcome of wider political economy of food, characterized by "trade liberalization and other state and suprastate regulatory structures and legal frameworks that favour large corporations, such as the food processors and agribusiness multinationals" (Otero *et al.* 2015: 8). Food security through economic access is not driven simply by sufficient money to buy food, but also by pricing structures that make some kinds of food more accessible than others. This is where economic accessibility meets nutritional outcome.

Other factors, seemingly unrelated to food, also have significant impacts on household food accessibility, such as instances in which households sacrifice economic access to food in order to meet other urban costs of living, such as the cost of transportation, rent, or water (PACSA 2017).

This chapter argues that it is important to expand the understanding of accessibility still further to reconnect to Sen's entitlement framework and consider inheritance- or transfer-based means of accessing food. In Sen's work these transfers referred to a far wider set of activities than rural–urban remittances, and yet most of the literature equates transfers with remittances. Recent studies and the findings of the Consuming Urban Poverty (CUP) project have indicated that the nature of urban and rural linkages is changing, and that remittances of food to urban households maybe becoming less important as a core source of food. However, inheritance- and intra-urban transfer-based means of access remain important.

The African Food Security Network (AFSUN) surveys of urban food security of low-income neighbourhoods in 11 cities in nine southern African counties found significant dependence on informal social safety nets to access food, with 41% of sampled households having shared meals with neighbours in the previous 12 months, and 21% having borrowed food (Frayne 2010). Work in Cape Town has found that a coping strategy for food-insecure households is to send children to eat at more food-secure households, who in turn respond by adding more "filler foods" (such as beans or potatoes) to the pot (Cooke 2012: 131). However, these social networks are fragile. In work in a community recovering from a fire in a low-income neighbourhood of Cape Town, Duncan (2013) found that individuals borrow strategically so as to not denude their social networks by over-extending good will. The careful negotiation of social networks as a means to access food was also noted by Davies's work in Lusaka (2016: 8): "Everyday acts of accessing and sharing food are embodied and enmeshed in relationships that are often unequal and fraught with power. Food activities are also bound up in the ability to have children maintain relationships of reciprocity and friendship."

Within the in-depth interviews conducted as part of the CUP project, the limitations of these social networks in a context of covariate scarcity were noted. Here interviewees spoke of adopting a number of outward-facing strategies to obtain food, including seeking food from non-governmental organizations, faith-based groups, and school meals, but were borrowing or sharing food within the community less frequently. Interviewees indicated that borrowing or sharing was only possible when neighbours or community members had surplus, and that in the current context of high food insecurity, dependence on these social networks was simply not possible. It is therefore essential to understand that accessibility is determined not just by household-scale poverty, but poverty at the community scale (Chileshe 2014).

Utilization

The FAO's utilization pillar correctly links the food system and possible food security outcomes to systemic issues beyond the food system. This chapter argues that this is particularly important in the urban context. However, while utilization is identified by the FAO as one of the four key pillars, its programming (and the national policies and programmes supported by it) continue to focus almost exclusively on availability or physical and economic (or even social) accessibility. Food utilization considers the ability of individuals and households to utilize food through "adequate diet, clean water, sanitation and health care to reach a state of nutritional well-being where all physiological needs are met" (FAO 2006).

This chapter argues that in order to understand the utilization dimension, it is essential to view urban food security as multi-scalar and multi-dimensional. At the household scale it is essential to consider the economic, infrastructural, social, and physical asset base of the household as it shapes food choice. While households may be physically, economically, or socially able to access "sufficient, safe and nutritious food which meets their dietary needs and food preferences for an active and healthy life," as per the FAO definition (FAO 1996), households still experience food insecurity due to a number of household-scale challenges. The first of these is the lack of adequate refrigeration and storage. Without these assets households are unable to safely store perishable foods, particularly in the context of extreme heat. As a result, households make a series of dietary choices to prevent spoilage, which impact the quality of diet and food security. First, they purchase perishable foods in smaller units (Ballantine *et al.* 2008). This means that they are unable to take advantage of the cheaper per unit cost benefits of buying in bulk. The urban poor pay more for fresh produce than the wealthy (Dodman *et al.* 2017). Second, a large proportion of food purchased that is to be consumed at home is processed foods that do not require refrigeration. These foods are often energy dense and micronutrient deficient, therefore resulting in less healthy diets and reduced ability to meet their food security needs.

A second set of asset constraints faced by poor urban households is the high cost of energy and water, which increase the cost of food preparation. Households have modified their diets as a result of these largely unaccounted-for costs of consumption. There has been a marked decline in the consumption of "traditional foods" (such as samp and beans in South Africa) that take a long time to cook (Drewnowski and Popkin 1997: 37), to consuming more processed foods that have a much shorter cooking time. This shift in diet cannot be attributed to energy costs alone but must be understood as the convergence of a number of factors. The first of these is the aspirational component of consumption, as people aspire to more "modern" diets (Noack and Pouw 2015). This includes both the actual foods purchased, but also the styles of cooking (fried instead of boiled, for example). The second is the increased availability of processed, quick cooking foods on the market. In South Africa, for example, sales of two-minute noodles ("ramen") increased by 40% between 2005 and 2010 (Igumbor *et al.* 2012: 1). In the CUP cities similar shifts were noted with *mandazi* (fried dough – Kenya) and *maputi* (popcorn – Zimbabwe) being sold by street vendors. Third, available food preparation time is increasingly limited in cities. Caballero (2005) suggests higher consumption of processed food may be the result of higher participation of women in the urban workforce, which limits food preparation time. A further important factor may be the longtravel times incurred daily by urban residents as a result of inadequate transport infrastructure and long distances between home and work.

The urban food system in African cities is responsive to these utilization challenges, and is also impacted by the same structural challenges. Small traders have responded to the lack of household storage and refrigeration by bulk-breaking products and selling food in small quantities. However, they are similarly affected by the limitations of storage and refrigeration, and therefore buy in relatively small units to ensure regular throughput of products and therefore food safety, as CUP research showed. The high cost of energy has provided an opportunity for street food vendors to produce pre-cooked foods for urban consumers unable to afford to buy raw materials and cooking fuel (Kimani-Murage *et al.* 2014). In Epworth, CUP project enumerators noted that street vendors had recently started cooking beans, packaging and freezing them, and selling them as a quick cook, low-energy dependent food. However, traders are also impacted by the high cost of energy, and a number of them use unsafe energy sources to operate (Dalvie *et al.* 2014). Finally, traders have responded to the time poverty of the urban poor by locating close to transport interchanges and selling prepared foods (Battersby *et al.* 2017). The CUP project's mapping of over 4000 food retailers in Kisumu, Kitwe, and Zambia found that traders were most active during later afternoon and evening commuting peaks, responding to the time poverty of urban residents.

Food safety is an essential component of utilization. As Duminy notes in Chapter 5, the dominant engagement of local governments with the food system has been through regulation for food safety. However, as the International Institute for Environment and Development (IIED) have illustrated, food safety issues emanating from food trade do not have their genesis in the traders themselves but rather in infrastructure inadequacies at the wider neighbourhood scale, including uncollected rubbish, and inadequate water and sanitation provision (Ahmed *et al.* 2014).

The outcome of this framing of food utilization is that it then requires that planning for food security extend beyond common household-scale interventions and suggests a wider set of potential interventions to improve food security. This framing therefore provides an opportunity for food security policy and programming to extend beyond the purview of Departments of Agriculture or Social Development and into a wider set of urban government departments.

Stability

The final FAO pillar of food security is stability. This is commonly a factor in rural food security in relation to crop seasonality. However, this pillar's definition encourages consideration of a wider set of determinants, as it states that

> they should not risk losing access to food as a consequence of sudden shocks (e.g. an economic or climatic crisis) or cyclical events (e.g. seasonal food insecurity). The concept of stability can therefore refer to both the availability and access dimensions of food security.
>
> (FAO 2006)

Work on urban food security has shown distinct temporal variations in food security. These temporal variations may be in part the result of food price fluctuations. Urban African households spend between 39% and 59% of their income on food, with poorer households spending an even higher proportion on food (Nakamura *et al.* 2016: 11). Poor urban households are very sensitive to food price fluctuations that may be seasonal. However, food price fluctuations are not the only drivers of temporal variations in food security.

There are distinct hungry seasons during the year (Chapters 16 and 17), and while food price changes play a role in these cycles, the hungry seasons are more closely related to the various urban costs of living. The hungry seasons are periods of increased household expenditure and reduced household income, as seasonal employment dips over these time periods. During these periods, households will decrease expenditure on food to meet other household needs.

They therefore "choose to go hungry" in order to preserve valued household assets (PACSA 2017). This again was evident in the CUP findings. The negotiation of food security is therefore a coping strategy that vulnerable households are able to draw on to mediate poverty. It is a coping strategy that has long-term negative repercussions, but it is also one of the few mechanisms by which poor households can control discretionary expenditure.

One means by which poor urban households are able to enhance their stability of food access is through accessing food from a range of sources. Food prices do not change uniformly, with different food items changing prices at different rates and at different times to others. Additionally, different retailers and retailer types respond to prices further down the supply chain differently. There are often considerable price differences between formal and informal sector retailers. Households will shop around to ensure they get the best prices. Another means by which households ensure stability of food security in the context of unstable livelihoods and fluctuating food prices is through participation in collective savings clubs, which can be drawn on in times of need. However, these are not viable for the most vulnerable who are unable to afford to invest in such savings groups (Davies 2016).

It is clear that to ensure stability of access to food for the urban poor it is beneficial to the urban poor to maintain a diversity of retail types, and to maintain non-market sources of food, including own production and social safety nets. It is important to note that the stability of different sources of food is impacted by external shocks and stresses that do not have uniform impacts on the different sources. Informal traders may therefore be more vulnerable to local environmental hazards and shocks, such as flooding, whilst large-scale formal retailers may be more vulnerable to other pressures, such as energy prices which affect the cost of doing business (Battersby 2012). This employment of diversification of food sourcing strategies to enhance stability therefore requires that policy makers and planners adopt an approach to food security that is based on an understanding of the food system that feeds the urban poor and on design interventions that support rather than inhibit food security.

Conclusion

The application of the four FAO pillars of food security to an understanding of urban food security in the sections above has illustrated the multiple ways in which poverty, as multi-dimensional and multi-scalar, intersects with food insecurity. Food insecurity is both a manifestation and a driver of poverty in African cities. Global discourses around food security have begun to change, with a shift away from a production focus towards a wider food systems approach (Lang and Barling 2012). As noted in Chapter 2, there has been a distinct and multi-programme shift in how food security is understood and governed, and how the challenge is engaged. The clearest sign in this shift is how the global governance institutions are engaging food security. Food security is no longer the domain only of the FAO. In the global research network, very different approaches are emerging. The World Bank, the International Food Policy Research Institute, and funders such as IFAD are all re-aligning their focus to consider the four pillars of food security but with two distinct differences. First, the production bias no longer dominates, with access and utilization attracting far greater attention. Second, the scale and locality of action is shifting, with the urban and city-regional scales increasing evident as areas of research, identified need, and governance intervention.

The new directions of research and action are demonstrating the potential richness of the four pillars in understanding the drivers of food insecurity and potential entry points for its alleviation. However, these pillars have generally been applied in an unbalanced manner with too great a focus on production and economic access. This chapter therefore calls for a refocusing and rebalancing of the four pillars to enable policy makers to develop urban policies and programmes in an integrated manner that engage food insecurity in all its dimensions.

This chapter has illustrated the multiple intersections between food, poverty, and the urban, and in doing so provides a refined understanding of urban food insecurity's drivers and potential points of policy, planning, and programmatic intervention. Responding to urban food security requires not just policy attention and appropriately scaled policy, but also consideration of a number of urban management activities, from health to urban planning, to transport and infrastructure planning. The multi-dimensional and multi-scalar nature of food security means that if these domains do not appreciate that food security is part of their mandate, the dimensions of food security will remain siloed and the impact of any intervention diluted.

Note

1 www.hivos.org/focal-area/sustainable-diets-all

References

Ahmed, S., Simiyu, E., Githiri, G., Acioly, A., Mbaka, S., Karanja, I. and Kigen, L. (2014, March). *Dining with less danger: Mapping food and environmental hazards in Mathare, Nairobi, IIED Policy Brief.* London: IIED. Available at: http://pubs.iied.org/pdfs/17218IIED.pdf [Accessed 22 Feb. 2018].

Badami, M.G. and Ramankutty, N. (2015). Urban agriculture and food security: A critique based on an assessment of urban land constraints. *Global Food Security*, 4, pp. 8–15.

Ballantine, N., Rousseau, G.G. and Venter, D.J.L. (2008). Purchasing behaviour as a determinant of food insecurity in Klipplaat. *Journal of Family Ecology and Consumer Sciences*, 36, pp. 1–8.

Barrett, C.B. (2010). Measuring food insecurity. *Science*, 327(5967), pp. 825–828.

Battersby, J. (2012). Urban food security and climate change: A system of flows. In: C. Moser, B. Frayne, and G. Ziervogel, eds., *Climate change, assets and food security in Southern African cities*. Abingdon: Earthscan, pp. 35–56.

Battersby, J., Marshak, M. and Mngqibisi, N. (2017). Mapping the invisible: The informal food economy of Cape Town, South Africa. *AFSUN, Working Paper Series24*, Cape Town,

South Africa: Available at: www.afsun.org/wp-content/uploads/2017/01/AFSUN24.pdf [Accessed 22 Feb. 2018].

Caballero, B. (2005). A nutrition paradox—underweight and obesity in developing countries. *New England Journal of Medicine*, 352(15), pp. 1514–1516.

Chileshe, M. (2014). *Economic shocks, poverty and household food insecurity in urban Zambia: An ethnographic account of Chingola*. Unpublished PhD, University of Cape Town.

Cooke, K. (2012). *Urban food access: A study of the lived experience of food access within a low-income community in Cape Town*. Unpublished MA Thesis, University of Cape Town.

Dalvie, M.A., Africa, A. and Naidoo, S. (2014). Relationship between firewood usage and urinary Cr, Cu and As in informal areas of Cape Town. *SAMJ: South African Medical Journal*, 104(1), pp. 61–64.

Davies, F. (2016). Lusaka's local food geographies: A gendered reading of everyday food insecurity in Mtendere, Lusaka. *Working Paper 40*. Cape Town: PLAAS, UWC and Centre of Excellence on Food.

Dodman, D., Leck, H., Rusca, M. and Colenbrander, S. (2017). African urbanisation and urbanism: Implications for risk accumulation and reduction. *International Journal of Disaster Risk Reduction*. Available at: http://dx.doi.org/10.1016/j.ijdrr.2017.06.029 [Accessed 22 Feb. 2018].

Drewnowski, A. and Popkin, B.M. (1997). The nutrition transition: New trends in the global diet. *Nutrition Reviews*, 55(2), pp. 31–43.

Duncan, S. (2013). *Food security in a post-fire disaster context: Experiences of female-headed households in an informal settlement*. Unpublished Honours Thesis, University of Cape Town.

FAO, (1996). Rome declaratin on food security and world food summit plan of action. *World Food Summit*, 13–17 November. Rome.

FAO, (2006). Food security. *Polic7 Brief June 2006, Issue 2*. Available at: www.fao.org/forestry/13128-0e6f36f27e0091055bec28ebe830f46b3.pdf [Accessed 22 Feb. 2018].

Frayne, B. (2010). Pathways of food: Mobility and food transfers in Southern African cities. *International Development Planning Review*, 32(3–4), pp. 291–310.

Global Nutrition Report, (2015a). *2015 Nutrition Region Review: Africa*. Available at: http://ebrary.ifpri.org/utils/getfile/collection/p15738coll2/id/130126/filename/130337.pdf [Accessed 22 Feb. 2018].

Global Nutrition Report, (2015b). *2015 nutrition sub-region review: Southern Africa*. Available at: http://ebrary.ifpri.org/utils/getfile/collection/p15738coll2/id/130151/filename/130362.pdf [Accessed 22 Feb. 2018].

Gómez, M.I., Barrett, C.B. and Raney, T. (2013). Postgreen revolution food systems and the triple burden of malnutrition. *Food Policy*, 42, pp. 129–138.

Harris, J., Drimie, S., Roopnaraine, T. and Covic, N. (2017). From coherence towards commitment: Changes and challenges in Zambia's nutrition policy environment. *Global Food Security*, 13, pp. 49–56.

HLPE (High Level Panel of Experts), (2014). *Food losses and waste in the context of sustainable food systems. A report by the High Level Panel of Experts on Food Security and Nutrition of the Committee on World Food Security*. Rome: HLPE.

Igumbor, E., Sanders, D., Puoane, T., Tsolekile, L., Schwarz, C., Purdy, C., Swart, R., Durão, S. and Hawkes, C. (2012). 'Big Food,' the consumer food environment, health, and the policy response in South Africa. *PLoS Medicine*, 9(7), p. e1001253.

Kimani-Murage, E.W., Schofield, L., Wekesah, F., Mohamed, S., Mberu, B., Ettarh, R. *et al.* (2014). Vulnerability to food insecurity in urban slums: Experiences from Nairobi, Kenya. *Journal of Urban Health: Bulletin of the New York Academy of Medicine*, 91(6), pp. 1098–1113.

Kirsten, J. (2012). The political economy of food price policy in South Africa. *Working Paper, 2012–102*. Helsinki: UNUWIDER.

Lang, T. and Barling, D. (2012). Food security and food sustainability: Reformulating the debate. *The Geographical Journal*, 178(4), pp. 313–326.

Maxwell, S. (1996). Food security: A post-modern perspective. *Food policy*, 21(2), pp. 155–170.

Nakamura, S., Harati, R., Lall, S.V., Dikhanov, Y., Hamadeh, N., Oliver, W.V., Rissanen, M.O. and Yamanaka, M. (2016). Is living in African cities expensive? *Working Paper 7641*. World Bank Group Policy Research, World Bank, Washington, DC.

Noack, A.L. and Pouw, N.R. (2015). A blind spot in food and nutrition security: Where culture and social change shape the local food plate. *Agriculture and Human Values*, 32(2), pp. 169–182.

Otero, G., Pechlaner, G., Liberman, G. and Gürcan, E.C. (2015). Food security and inequality: Measuring the risk of exposure to the neoliberal diet. *Simons Papers in Security and Development, no. 42/2015*. School for International Studies, Simon Fraser University, Vancouver, March.

PACSA, (2017). *PACSA 2017 Food Price Barometer: Annual Report*. October 2017. PACSA, Pietermaritzburg. Available at: www.pacsa.org.za/images/food_barometer/2017/2017%20PACSA%20Food%20Price%20Barometer%20annual%20report.pdf [Accessed 22 Feb. 2018].

Rose, D., Luckett, B. and Mundorf, A. (2013). *Diet Matters: Approaches and Indicators to Assess the Role of Agriculture in Nutrition*. Rome: Food and Agriculture Organization of the United Nations.

SACN, (2015). *A study of current and future realities for urban food security in South Africa*. Johannesburg: South African Cities Network.

Sen, A. (1981). Poverty and famines: An essay on entitlement and deprivation. Oxford: Clarendon Press.

United Nations General Assembly, (2015). *Transforming our world: The 2030 agenda for sustainable development*. A/RES/70/1, 21 October. New York, NY: United Nations, Department of Economic and Social Affairs.

Warren, E., Hawkesworth, S. and Knai, C. (2015). Investigating the association between urban agriculture and food security, dietary diversity, and nutritional status: A systematic literature review. *Food policy*, 53, pp. 54–66.

Weingartner, L. (2004). *The concept of food and nutrition security*. Background paper #1: Food and Nutrition Security Assessment Instruments and Intervention Strategies. Avalaibale from: http://www.oda-alc.org/documentos/1341934899.pdf [Accessed XXX].

WHO/FAO, (2002). *Diet, nutrition and the prevention of chronic diseases: Report of a joint WHO/FAO expert consultation*. Geneva: WHO.

Wiggins, S. and Keats, S. (2015). *The rising cost of a healthy diet: Changing relative prices of food in high-income and emerging economies*. ODI Report May 2015. London: ODI.

4 Understanding and addressing poverty, labour force, and urbanisation data gaps in sub-Saharan Africa

Jacqueline Borel-Saladin, Muna Shifa, and Ann Donald

Introduction

The major outcome of the United Nations Conference on Housing and Sustainable Urban Development (Habitat III) in October 2016 in Quito was the New Urban Agenda (NUA). The NUA is a "global commitment to sustainable urban development" and is viewed as a "critical step for realizing sustainable development" (United Nations 2016: 3). It is also integral to the achievement of the 2030 Agenda for Sustainable Development, especially Goal 11 of "making cities . . . inclusive, safe, resilient and sustainable" (United Nations 2016: 4). This document also recognized the crucial role of data in the successful implementation of the NUA, particularly the potential ability of data to strengthen and support innovative science and research in urban planning and policy formation; the monitoring of progress towards the Sustainable Development Goals (SDGs) and informed decision making; the role of all levels of government in data collection, mapping, analysis, and dissemination; evidence-based governance, based on censuses, household surveys, population registers, and community-based monitoring processes; and the promotion and enhancement of open data.

The importance of data in the NUA was reinforced in the Cape Town Global Action Plan for Sustainable Development Data, drafted at the inaugural United Nations World Data Forum in January 2017. It emphasizes specifically that "[statistical] capacity building is important for all countries, even more so for developing countries, particularly African countries" (HLG-PCCB 2017: 1). The conclusion states that the over-arching goal of this plan is to "address gaps in national statistics and statistical coordination . . . not restricted to capacity building for SDG indicators alone [but] to strengthen the national statistical systems so that they can be most responsive to statistical needs to achieve the 2030 Agenda and beyond" (HLG-PCCB 2017: 8).

In this chapter, we reflect on these data goals considering the specific difficulties faced by researchers in the Consuming Urban Poverty (CUP) project. The project aimed to examine the dynamics of urbanisation, poverty, and governance in secondary cities in Africa with a focus on two secondary cites (Kisumu in Kenya and Kitwe in Zambia) and a satellite settlement of a capital city (Epworth in Zimbabwe). Certain data were required for particular lines of inquiry in this project, namely: urban poverty and inequality (ideally by means of income and expenditure measures, the construction of multidimensional poverty indices, and the generation of small area poverty estimates); urban labour market change (through changing employment patterns over time, via occupation and industrial sector statistics); and urbanisation (using GIS to analyze land use and land cover change). Data on these topics exist at a national level for all three of the countries in the CUP project; however, our focus is not at the country scale, but rather the city scale. Data at metropole or primate city level would potentially not be too difficult to source, but our focus was on the secondary cities of Kisumu, Kitwe, and Epworth, the data for which is considerably more difficult to source. As such, in this chapter we will discuss the unique challenges that came with acquiring the necessary data for studying poverty, labour market change and urbanisation in these smaller African cities. Figure 4.1 below sets out the five key data challenges faced in this project and the questions which were posed in relation to each of them. The chapter sections discuss each of these in turn. We conclude by offering possible solutions to the problem of the lack of more suitable, readily available data to address these data gaps.

Data availability and access

Despite the fact that plenty of data have been generated in African countries, data gaps persist as a result of issues of access to them. Open data is the "first prize" for researchers, as it is often freely available and easily accessible via the internet. Kenya is the only country in the CUP project with a functioning national data archive accessible through an online data portal. One of the Kenya National Bureau of Statistics' (KNBS) core mandates is the dissemination of statistical information and it has had a policy for the release of anonymized microdata files since approximately 2008 (KNBS 2008). After registering on this system, researchers can access multiple data sets generated from censuses and various household surveys on a wide variety of topics, e.g. health, remittances, enterprises, food, household budgets, housing, agriculture, the labour force, health, and welfare.

Even in countries without clear data dissemination policies, there are often official channels through which data can be obtained from national statistical offices. A letter of request can be sent to the director of such a body stating what data are required and how they will be utilized. This approach has certain disadvantages though. First, while the recipients of the data do not have to be physically present in the country from which the data are being requested, this process does tend to require a representative of some sort to meet face-to-face with the director general to discuss the request. This process creates delays. Furthermore, overcoming this obstacle is no guarantee of actually receiving the requested data either. In the case of CUP, access was granted to census and various survey data and poverty statistics from both the Central Statistical Office of Zambia and the Zimbabwe National Statistics Agency (Zimstat). However,

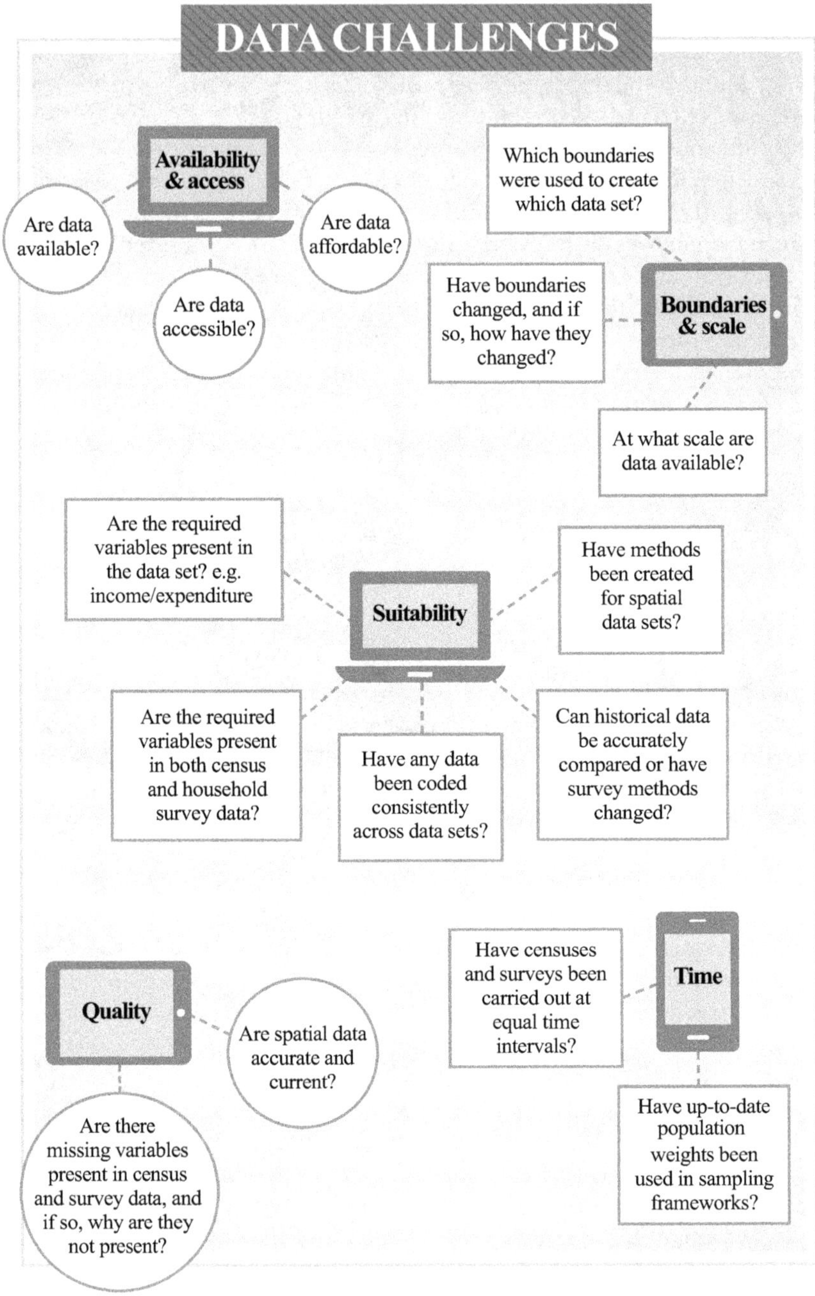

Figure 4.1 Organizational chart of data challenges

(Author's own)

data were only received once money was sent to the contacts in each of these statistical offices.[1] These types of costs associated with the purchasing of data can be a major hindrance to researchers working on less-well funded projects.

Once permission for access to data had been granted in Zimbabwe, Zimstat employees generated specified cross-tabulations for researchers. This was highly problematic as it severely impacted the quality and suitability of the data (see sections on suitability and quality). The reason provided for this limited access to microdata was the fear that confidentiality protection might somehow be jeopardized. This is, however, somewhat confounding as it is for this exact reason that data are anonymized – a process in which the identifying variables contained in the data set are modified to protect individuals' privacy (Trewin *et al.* 2007). Direct identifier variables such as names, addresses, or identity numbers do not appear in final data sets. Respondents are usually represented as a numbered household member with a household code. In various tests performed by skilled researchers working in co-operation with national statistics offices, and with substantial financial and technical resources, it has been shown that chances of identification of individuals in official data sets are often overestimated (Dale and Elliot 2001). Their conclusions suggest that it would be relatively difficult for someone without access to multiple other pieces of relevant personal information to identify an individual from a census or survey.

Ownership, licensing, and cost issues arise with spatial data too. Government departments may not always be willing or allowed to provide easy access to data. Municipal departments also often have various spatial data sets which are not always freely shared between departments (Baud *et al.* 2014). Developing countries face additional challenges in generating and sharing spatial data such as limited budgets to cover the costs of the necessary hardware, software, and images, a lack of specialist capacity, and slower internet access. Data do not have to be state-generated and centrally controlled though. It can also have been gathered by private companies; however, many companies do not want to make commercial data freely available and purchasing such data sets may be prohibitively expensive for researchers.[2] New participatory approaches to data generation are leading to community generated/collected data being freely available. For example, under the Missing Maps project, remote volunteers trace satellite imagery into OpenStreetMap, which is then populated with details such as neighbourhood boundaries and street names by local volunteers "on the ground" (www.missingmaps.org). All data gathered are freely and openly available under the project's open license. In the case of Kitwe and Kisumu, incomplete data sets are available on OpenStreetMap. The study site of Epworth, Harare, was mapped as part of this project from March 2015 onwards and a validated spatial data set has been created from this.

One aspect that all these ways of finding data have in common is the use of a local informant or key person. These can take the form of local researchers, employees of various state departments (such as national statistical or town planning officers), members of NGOs operating in a specific location, and "ordinary" citizens. These people have valuable local knowledge of where to find specific types of data or whom to approach for it, as well as being able to provide access to data as they themselves may act as gatekeepers to it.

Unfortunately, some data simply are not publicly available. In terms of spatial data, this is usually related to licensing and ownership issues. However, there are plenty of data sets generated through research using public money that are not made available to the public. For example, one approach for analyzing poverty at lower geographical units such as secondary cities and smaller towns is to use small area poverty estimation, which is called "poverty mapping." This combines household survey data on expenditure or income with census data. However, full census data sets are required for this and usually only the 10% sample sets are available. Full census data sets are very big files, and downloading them online from a data archive, for example, can be difficult (if not impossible) if there are internet connection problems. This is a very real obstacle in many parts of Africa. Furthermore, sometimes no data sets, not even sub-samples, are available.

Boundaries and scale

After the necessary types of data sets have been acquired, the next major problem when studying any area is around geographical boundaries. Boundaries often are not determined by natural physical features that are stable and unchanging over several decades. Usually they are man-made, reflecting political or administrative considerations, and can be remarkably changeable over relatively short periods of time.[3] Kenya is a prime example of the vulnerability of area boundaries to political whims. Kisumu district only came into being in 1964. In 1998, a new district (Nyando) was split off from the original Kisumu district. Around 2007, Kisumu was split into Kisumu East and West districts. These were part of a flurry of district creations that occurred after 1992, in which approximately 265 additional districts were created. Additional districts were created ostensibly to bring government services such as health, security, courts, schools, land transactions and disputes, and financial services closer to where they were needed. New districts were mostly announced at political rallies though, without any prior official planning, and there was never any real infrastructure or financing put in place to support them (Nguru 2012). In 2010, the courts ruled that any districts created after 1992 were done so "in complete disregard for the law," and therefore illegal. The provincial administrative offices continued to use these districts though, supposedly to ease the transition to the new system. The result is that different data sets have different districts, and when combined may or may not reflect the currently accepted boundaries of the newly formed Kisumu County.[4] This makes it very difficult to accurately identify Kisumu over time in both survey and spatial data. In theory, any change regarding location boundaries should be corrected in each data set; however, these frequent location boundary changes complicate these corrections, and as such often very little systematic documentation of these changes takes place. Such inconsistencies in geographic location identification codes in the census and household survey data have a major impact on the accuracy of estimates and results.

The identification of larger districts is particularly salient in the study of smaller African cities, as districts often have to be used as a proxy for the city because of the scale of the survey. In most household surveys, detailed data are only representative at the national, provincial, and district scales. This is not a problem when studying the capital city of an African country, as it is often large enough to constitute its own district; however, this is a problem in the case of smaller cities and towns. This is not to say that detailed data for poverty measurements and labour market change are not collected for individual cities and towns. It is rather an issue caused by the fact that the sampling frameworks for these surveys are only designed to be representative at larger scales. As such, the samples of individuals in individual cities/towns, even those with relatively large populations (e.g. secondary cities), are too small to be representative of the whole population of that city/town and give unbiased estimates. Thus, it is not possible to generate accurate poverty, inequality, and labour force estimates at lower levels of geographic aggregation (secondary cities or small towns) using only household surveys.

In other cases, the greater district is used as a proxy for the city because the relevant spatial location variable that would enable one to identify a city or town has been omitted from the publicly available data set. For example, the population census collects data and is representative at not only the larger provincial and district scales, but also the smaller administrative scales such as city/town, division, location, and ward.[5] However, publicly available censuses for both Zambia and Kenya do not include a geographic location code variable for the identification of secondary city boundaries. In Kenya city/town code identifiers were collected in the 2009 population census data, but are not supplied to researchers without a lengthy request for special permission. Presumably, this is again related to concerns about protecting individual's identities. However, removing entire records is an extreme measure of data protection (UNECE 2007). The 2009 Kenyan census includes spatial location variables at an even smaller scale than city/town, including location, sub-location, and enumerator area, and is freely available to anyone to download from the data archive site as an Excel spreadsheet with the corresponding names for these codes. This, in conjunction with the standard practice of anonymizing the data, makes it unclear how not having a city/town name variable in such a data set helps protect individuals' privacy.

Spatial data are no less subject to considerations of scale. The size of the pixel captured by a satellite (resolution) determines how much detail can be seen in the resulting image and whether or not specific features of the urban landscape can be differentiated. Only higher resolutions can produce the detailed imagery required to discern change in smaller objects in urban landscapes, such as buildings and houses. Images with lower resolutions can be used to simply monitor changes in the extent of an urban area over time. However, urban areas are very heterogeneous and as such, higher resolution data are required to accurately map them. In developing world countries in particular, informal settlements are growing in size and number as urbanisation increases. Isolating this in satellite imagery requires higher resolutions to identify individual buildings in unstructured, densely occupied slums. Higher resolution imagery also enables the creation of detailed base maps which can be used in further studies. Thus, higher resolution imagery is all around more useful for effectively studying land use and land cover change in urban areas in Africa; however, higher resolution imagery is also costlier and thus accurately capturing the urban landscape requires large capital investments.

Suitability

Even if data sets sampled at the appropriate scale with correctly adjusted boundaries can be obtained, they do not necessarily contain the variables that are needed for specific types of research. Although census data can give us representative samples at lower geographic units, income poverty often cannot be measured using census data as most censuses do not collect information on income or expenditure. This is a serious limitation, as income is a key dimension of poverty in urban settings as the urban economy is highly monetized – access to food, shelter, water, sanitation, etc., all involve some form of payment.

According to the "small area poverty estimates" approach, we can use the survey data to predict consumption, then use the parameter estimates from this exercise to predict consumption/expenditure in the census (using similar variables from the census as those used to predict consumption from the household survey). Turning to this approach as an alternative does not necessarily help, however, as the variables used to predict household consumption have to be limited to only those variables that can be found in both the census and household surveys, as the lack of a larger number of comparable variables used to predict income and expenditure in the census and surveys diminishes the accuracy of the poverty estimates. A similar problem exists with constructing multidimensional poverty estimates. More recent censuses include many indicators of welfare measures, such as access to various basic services and asset ownership (both household durables and livestock holdings), education of each household member, and housing characteristics. Previous censuses either do not include several of these indicators or, if they do, the variables are not coded consistently across different census years. Therefore, an analysis of progress in multidimensional poverty can only be based on a restricted number of indicators, which could be misleading. Household surveys that collect information on consumption also do not provide comparable estimates over time. For example, major changes in Zambia's Living Conditions Monitoring Surveys questionnaire design saw the list of consumption items increased from just 87 in 2006 to 213 in 2010. The upshot of this is that the consumption aggregates calculated for these two surveys are not comparable, neither to each other nor to earlier surveys. Careful adjustments need to be made to these data if they are to be useful in poverty estimation. This lack of consistent poverty estimates over time is a serious limitation when analyzing the relationship between urbanisation and poverty trends.

Lack of consistency over time is also a major limiting factor in tracking labour market change. Ideally, a survey or census uses the International Standard Classification of Occupations (ISCO), which is list of occupations, classified at various levels of detail, designed expressly to facilitate international comparison (ILO 2012). If this system is not used, it is very difficult, if not impossible, to compare occupation and employment sector variables over different surveys and censuses. One possible solution to this problem is to merge certain occupation categories together, meaning there will be fewer and less well-defined occupation groups that are comparable over different data sets. However, often in doing this, so much detail is lost in aggregating the original data into the resultant macro occupation groups that they become essentially meaningless.

Consistency is also an important determinant in spatial data. All data are created with similar naming conventions, levels of accuracy, data formats, and standard metadata. This allows easy interpretation, use and comparability. Standardization is fundamental to spatial data. Data sets must be interoperable, i.e. they can be used on different computer systems and in different software packages. Standardized metadata systems are also of particular importance as they enable users to easily access information such as data creation processes and the coordinate systems used. Without these standardized details, it is very difficult to effectively utilize spatial data in any study. Secondary spatial data obtained from government and/or private organizations may not always have standard metadata applied, making it harder to accurately interpret. In the case of OpenStreetMap, the approaches to capturing data are highly standardized and publicly documented, enabling effective use of the data.

In contrast to the benefits of standardization in spatial data, applying standardized poverty measures to different contexts undermines the accuracy of poverty estimates. A single urban poverty line is used to estimate income poverty due to the lack of systematic correction for differences in the cost of living across different urban categories (price data). This is applied to all urban areas, be they capital cities, secondary cities, or smaller cities and towns. Yet costs of household and other services vary across different urban areas within a given country. The suitability of the available indicators to measure urban poverty is also a key limitation when considering non-income dimensions of poverty. There is a limited selection of poverty indicators that are urban specific. Also, in the CUP project, the same deprivation indicators had to be used for both rural and urban residents in the calculation of multidimensional poverty; however, deprivation and poverty are context dependent. Most of the indicators also do not reflect differences in quality of services in different places, e.g. the quality, frequency, or affordability of access to public services, such as water and electricity, and quality of housings in urban areas.

There are very different issues in the case where no raw data are available and cross-tabulations are generated by statistics office employees. For poverty estimates, cross-tabulation results are of no use. Only the microdata, with data for each individual sampled rather than the aggregated results of a cross-tabulation, can be used for statistical tests and modelling. Cross-tabulations of census data can be used for studying labour markets in terms of monitoring changes in occupation and sector of employment (descriptive statistics).

Quality

Missing variables in available data sets severely impacts the quality of these data and limits their utility. The 2010 publicly available Zambian census from IPUMS does not include a variable to identify rural and urban areas. Given this, it is difficult to estimate poverty in urban Zambia, let alone to talk about poverty in secondary cities. Other census data obtained from the Central Statistical Office of Zambia does not include a unique household identifier variable. This makes it impossible to merge the individual-level data with household-level data so that all the characteristics required for calculating poverty estimates can be assigned to each respondent. Are these variables omitted to protect individuals or is this an oversight? It seems counterproductive and wasteful to go to the effort and expense of collecting data at smaller spatial scales but then purposely omit the variables that would allow them to be used to study these smaller units.

The quality of the cross-tabulation results received from statistics officers is also a concern. Instructions can be given by a researcher as to how to generate the desired cross-tabulations, but one has no way of knowing what procedures have been followed on receiving the output. Even when giving instructions, the results received contain inconsistencies and/or do not appear feasible, making them questionable.

Spatial data quality problems include the data not being current and up to date, in the correct formats or projections, or available at the necessary scale. Often data vary markedly in their accuracy, precision, completeness, and consistency. In addition to this, any user of a data set should be able to glean the following from the metadata: who created it, when was it created, why it was created, how often/when it was last updated, what the accuracy levels are, and what kind of data are used. These metadata may be missing entirely or be incomplete. Different areas have differing amounts of coverage too. Another problem is that there can also often be various versions of the same data across different government departments.

Time

While the geographical scale of the data is the major constraining factor when studying smaller cities, the time scale is no less important, with census data losing its validity over time (Baud *et al.* 2014). Long periods often lapse between successive waves of censuses or surveys. This is a key limitation in using censuses or other low frequency data to monitor progress over short time periods, e.g. to examine the impacts of shocks and seasonality. Censuses collect information on basic non-income dimensions of welfare indicators only every ten years in many countries. This problem is not limited to census data only though. For example, in Kenya, prior to the 2005/06 Kenya Integrated Household Budget Survey, the most recent household survey that collected detailed expenditure data required for poverty measurement was the 1997 Welfare Monitoring Survey. Also, there is often a time-lag between censuses and household surveys and the sampling frameworks are not updated to reflect the current population. Population weights calculated from previous censuses are also used in other data sets when more recent census population weights should be used. The limited number of comparable data sets in some cases also make it difficult to study trends over time.

In spatial analyses examining changing land use and cover over time, data selection is heavily influenced by what is available historically for a particular area and which sensors were in use over the relevant time period. Medium resolution imagery such as Landsat and SPOT can offer a sufficient balance of spatial detail and availability of historical images. In the CUP project, high-resolution imagery was obtained for 2015, but earlier imagery is only available in lower resolutions. Some of the oldest imagery for some of the cities enables the extraction of city boundaries only, with no identifiable internal urban detail. Reconciling imagery of decreasing resolution as one goes further back in time poses a considerable challenge in this kind of study of urban change.

Conclusion

Ideally researchers should have free and open access to all household survey and census microdata sets and spatial data. Depending on the kind of research being conducted, unabridged microdata sets, for example, are a necessity, e.g. for the calculation of poverty estimates and statistical modelling. In this project, very little data were accessed in this manner. Even when working through statistical offices and official processes, payment still had to be made to receive data. This led to serious delays. Unfortunately, in the case of Zimbabwe, census and survey microdata sets are not made available at all, severely limiting the research team's work. Comparability over time is a major challenge, and missing and inconsistent variables between different waves of the same survey make it difficult to calculate poverty statistics and study labour force changes. Inconsistency in the availability of specific resolution spatial data also makes mapping urbanisation difficult.

More regular censuses (not only every ten years) and household surveys reflecting incomes, expenditure, assets, detailed migration patterns, and occupation profiles are needed in order to better study poverty, inequality, labour markets, and urbanisation in secondary African cities. In large part, the survey and spatial data that researchers need already exist in various forms in national statistical offices and other government departments, e.g. town planners. However, there are serious problems of data accessibility to entire data sets or key variables, often justified in terms of protecting privacy. By making access to already-collected data easier, considerable time could be saved and the analyses requiring microdata greatly improved. In many African countries, there are serious technical and financial obstacles to the appropriate handling of the more complex aspects of data management. Additional funding, be it from the national fiscus or donor organizations, is most needed for capacity development, both in terms of technical skill and equipment. The legislation around statistics and data are outdated in many countries too, especially in terms of the dissemination of electronic data. This is perhaps reflective of a tendency in some places to control access to data because of concerns over its misinterpretation or fears of abuse. However, this legislation needs to be updated in accordance with the latest international protocols and best practice. This can help create a more open data landscape where researchers are not at the mercy of gatekeepers and can timeously obtain the high-quality data they need for valuable research that contributes to our understanding of African cities. The NUA outlines ambitious goals for global urban development, but also recognizes the unique challenges facing developing countries. Given the data requirements to help reach these goals, the kinds of problems discussed here make this particularly challenging in Africa.

Notes

1 In the case of Zambia, this was ostensibly for the cost of couriering disks of data to South Africa. In the case of Zimbabwe, no reason was given for why money was needed.
2 Commercial companies may also be generating the data on behalf of less well-equipped and experienced government departments, which could pose problems around data licensing and ownership.
3 Changing boundaries or urban areas do not, however, affect the spatial analysis, as they are man-made boundaries with no real presence in the natural landscape.
4 In conformity with the new constitution enacted on August 27, which created counties as the focal administrative/political units (KNBS, 2012).
5 The population census is supposed to include the entire population. Theoretically sample sizes are not an issue.

References

Baud, I., Pfeffer, K., Scott, D., Denis, E. and Sydenstricker, J. (2014). Participatory 'Spatial' knowledge management configurations in metropolitan governance networks for sustainable development. *Chance2Sustain: Work Package 5 Thematic Report*. Bonn: EADI. Available at: www.chance2sustain.eu/fileadmin/Website/Dokumente/Dokumente/Publications/publications_2014/C2S_TR_No02_WP5_V5-6.pdf [Accessed 23 Feb. 2018].

High-level Group for Partnership, Coordination and Capacity-Building (HLG-PCCB), (2017). *Cape Town global action plan for sustainable development data. 15 January 2017, Cape Town*. New York, NY: United Nations Statistical Commission.

Kenya National Bureau of Statistics (KNBS), (2008). *Data access and dissemination policy*. Nairobi: KNBS.

Kenya National Bureau of Statistics, (2012). *Kenya 2009 population and housing census analytical report on Kenya population Atlas*, Vol. XV. Nairobi: KNBS.

International Labour Office (ILO), (2012). *International standard classification of occupations (ISCO-08): Structure, group definitions and correspondence tables*. Geneva: ILO.

Nguru, S. (2012). *The 265 Districts of 2010*. Available at: www.ustawi.info.ke/index.php/devolution/265-districts-of-2010 [Accessed 12 Jan. 2018].

Trewin, D., Andersen, A., Beridze, T., Biggeri, L., Fellegi, I. and Toczynski, T., (2007) *Managing statistical confidentiality and microdata access: Principles and guidelines of good practice*. Geneva, UNECE/CES.

United Nations General Assembly, (2016). *Draft outcome document of the United Nations Conference on Housing and Sustainable Urban Development (Habitat III)*. 17–20 October, Quito. New York, NY: United Nations.

Part II

Urban food governance and planning

5 Historical urban food governance in Africa

The case of Kenya, c. 1900 to 1950

James Duminy

Introduction

Using the specific case of Kenya, this chapter explores how urban food issues emerged as problems of colonial government in Africa. The objective is to give historical context to one of the main arguments pursued by this book: that little attention has been paid to the local governance and planning of urban food systems in Africa, and that food policies pursued by central governments have tended to focus on increasing food availability and rural agricultural production. In order to understand why local governance has engaged so little with systemic food issues, we need to understand more precisely, on one hand, how urban governance *has* interfaced with food problems and, on the other, how food governance has responded to urban realities and risks.

Most of the existing historical work on urban food governance in Africa has focused on how urban food supply and marketing systems functioned in various contexts (Guyer 1987a; Porter *et al.* 2007). Relatively little attention, however, has been directed at official responses to urban food issues in relation to problems such as disease, nutrition, poverty, and labour unrest (Clayton and Savage 1974; Cooper 1987). Work on the history of African colonial town planning points to official interest in controlling disease and migration and promoting racial segregation (Home 2012; Njoh 2009; Silva 2015), but has not examined the relationship between urban governance and food problems in significant detail. As such, we have little sense of why a gap, in thought and practice, exists between food and local urban governance in Africa. Here I seek to provide an initial historical sketch that may help in starting to traverse this gap.

The empirical focus is on Kenya in the first half of the twentieth century. The concentration on Kenya, as a case to illustrate wider historical dynamics in Africa, is justifiable for two reasons. First, colonial Kenya is a context that is relatively under-examined in the literature on African food and famine history. Second, as neither the wealthiest or poorest African colony, Kenya exemplifies many of the trends and challenges faced by officials in settler colonial contexts of East and Southern Africa (Gardner 2012). Where possible, a comparative perspective with other colonial African contexts has been incorporated.

The argument presented in this chapter is as follows. For the initial period of colonial rule in Kenya, at the start of the twentieth century, the foremost problem facing officials was rural scarcity. Their responses generally focused on enhancing or increasing agricultural production. Colonial states, like that of Kenya, did attempt to regulate urban food systems, albeit in different ways, at different times, and at various scales. In Kenya, the main features of the state's mode of understanding and managing urban food problems developed during the interwar period and were established by the end of the Second World War. In this time, we see the emergence of a dual mode of addressing the urban food system. This manifested, on one hand, in concerns over food contamination and the spread of disease, for which official powers were secured over the production and use of the urban built environment. On the other hand, we find anxiety over urban food supply and nutrition, understood as a wider economic problem involving food production, distribution, pricing, and income. As such, in colonial Africa the role of local urban governance in relation to food systems was limited to regulating and preventing certain kinds of food preparation, supply, and trade. Local government has had little influence over the production, distribution, and pricing of food supplies – this being the domain of a centralized statutory marketing system that was established by 1945.

The chapter proceeds in three parts. The first describes how rural food scarcity emerged as a principal problem for colonial officials over the first four decades of the twentieth century. The second outlines how colonial officials responded to urban food problems in the domain of public health. The third section discusses how the nature of urban food problems shifted during the Second World War, in the process establishing an economic view of food access and nutrition.

Scarcity and famine

The early years of colonial rule in Kenya encompassed several periods of acute food scarcity – the first being the "Great Famine" of 1898 to 1901, which led to thousands of deaths and massive population displacements (Iliffe 2007). The colonial administration's role in confronting famine events was limited to providing emergency relief, either in the form of free or subsidized supplies for the destitute (prioritizing women, children, and the disabled), or by arranging relief works where able-bodied men could labour in exchange for food supplies. Officials also used the opportunity of scarcity to "encourage" African people to seek wage labour on settler estates as a means to secure both cash and food rations. Kenyan administrators mostly relied upon private traders to move food supplies to the eager market of the hungry during times of dearth, as did their counterparts in Southern Rhodesia (Iliffe 1990). In part, this reliance on private trade reflected the colonial state's incapacity to conduct large-scale relief operations by its own means. Yet it should also be seen against the general backdrop of a longstanding British ideological aversion to state intervention in food pricing and distribution, which had taken firm hold of official opinion in the nineteenth century (Davis 2001; Vernon 2007).

The anti-famine measures pursued by the colonial state reflected the conviction that African communities and regions were, and should be, self-sufficient in food (Bryceson 1988; Carswell 2003). In Kenya, the ideology of regional self-sufficiency was most visibly expressed by the 1922 Native Foodstuffs Ordinance. This legislation enabled officials to restrict the exchange and export of vital African foodstuffs from any district suffering from food shortages. Influenced by similar legislation developed in Nyasaland and Uganda, the Ordinance coincided with a short-lived and half-hearted official endorsement of African export production in the early 1920s (Van Zwanenberg and King 1975). While mostly applied to rural districts, officials also used the legislation to protect the food supplies of growing urban centres like Nairobi (Robertson 1997). The practice of controlling the inter-district movement and sale of goods later formed the basis of the 1935 Marketing of Native Produce Ordinance, which granted the governor powers to declare areas in which the purchase and sale of African produce (particularly maize) would be limited, controlled, and regulated. By the mid-1930s, government policy had shifted decidedly towards encouraging African export production. This tied into an overall anti-famine strategy that sought the production of an exportable surplus of maize as insurance against local scarcities (CPK 1943).

Ordinances such as these demonstrated the growing political will to intervene in the agricultural and food marketing system in the interwar period (Gardner 2012). This willingness was driven by different motives, and often emerged in response to specific crises. The official response to the Kenyan famine of 1929–1930 is a key example of this interventionist trend. Here a prolonged drought and widespread locust damage led to acute scarcities in many rural areas, and particularly in the Meru and Embu Districts. Fearing an all-out famine, in early 1929 officials passed a Food Control Ordinance and hurriedly formed a central Food Control Board. The Board was empowered to collect statistics on available foodstuffs, grant licenses for trade in food, acquire and distribute supplies (purchased from settler farmers) for famine relief, requisition transport at set rates, fix maximum food prices, and if necessary import additional supplies to supplement local stocks.

On one level, the measures contained within the Food Control Ordinance reflected a paternalistic notion of the state's duty to save starving African subjects (Vernon 2007). On another, it demonstrated officials' eagerness to avoid the consequences of food shortages and price inflation. In particular, they worried that rampant inflation of maize prices (aggravated by "unscrupulous" traders and middle-people) would hurt the economic position of both private enterprise and the state (Holm 1929). Their concern was that high maize prices would boost the production costs of Kenya's "consuming interests" (notably large employers of labour such as rural sisal plantations, as well as the railway and government departments, with their predominantly urban labour forces) to the point where profitable production or administrative solvency were under threat. Food price inflation also affected urban populations who, while unlikely to starve during times of dearth, were more dependent on cash food purchases and thus prone to experience scarcity in the form of reduced real wages and malnutrition.

For the first three decades of the twentieth century, then, the main food-related problem energizing officials was that of rural scarcity and hunger. Officials faced the problem of distributing emergency supplies to remote and marginal rural areas, where people were most vulnerable to disruptions in climate and normal trading relationships (Iliffe 1990). Yet industrial and urban growth were changing the nature of food scarcities and the economic risks they posed to consumers and employers. This trend would come into sharp focus during the Second World War – a subject to which I return in the sections that follow.

During the 1920s and '30s, food scarcity was not a specifically *urban* problem primarily because of scale. Colonial African towns were generally administrative centres with relatively small permanent populations. In any case, officials could always resort to more coercive practices of population registration, migration control, and repatriation – responses linked to fears around African detribalization and delinquency resulting from rural–urban migration (Campbell 2002) – to restrict urban food demand and alleviate the potential political consequences of scarcity (Guyer 1987b). At this stage, the concerns of state and municipal administrators lay more with another aspect of the urban food system: that of public health.

Food and urban public health

In Kenya, official interest in urban food systems and public health manifested in two main areas. First, in regulating the production and use of space to improve sanitation and prevent outbreaks of disease. And second, from the late 1920s, in efforts to measure and regulate the cost of living and nutritional status of urban populations, particularly African labour.

While rural administrators were preoccupied with issues of scarcity, the major food-related problems confronting those in charge of urban areas were rather different. In the early twentieth century, food was drawn within broader colonial anxieties and regulations surrounding urban public health. Generally, Kenyan officials worried about the threats of communicable disease presented by the poor sanitary conditions prevailing in the colony's growing towns (EAP 1910). Some such diseases, including typhoid fever, dysentery, and cholera, were able to spread through contaminated food and water supplies. Accordingly, to reduce the risk of contamination, officials sought to regulate places where food was prepared, stored, and sold (including markets, abattoirs, shops, and "eating houses"); the individuals involved in preparing, moving, or selling food; and sources of food and food articles themselves. Food thus provided a key part of the justification for the colonial state to intervene in the production and use of public and private urban space, including (ultimately unsuccessful) public health arguments for the entrenchment of urban racial segregation as official state policy (Greenwood and Topiwala 2015).

In Kenya, for the first three decades of colonial rule, public health administration in the towns was the domain and responsibility of the central government. Urban health officials and inspectors were employed directly by the state's Medical Department, which acted as an independent executive authority in relation to local government bodies. State actors thought it imperative to devise a universal public health policy, applicable throughout the territory, as sanitary and health matters, while concentrated in the towns, held such importance for the general welfare of the country (CPK 1927). In other words, the risks of urban disease outbreaks were seen as epidemical threats to the colony as a whole, calling for direct, preventative intervention and overall coordination by the state. Centralized control was further justified by the relatively inchoate nature and capacity of colonial local government in the early twentieth century. Until the late 1920s, all Kenyan towns (excepting the capital, Nairobi) were administered directly by the central government under the 1903 Townships Ordinance, with local district officers sometimes assisted by resident advisory committees (CPK 1929). Nairobi, by contrast, was from 1919 governed by a municipal council that provided local services such as water supply, sewerage, and drainage, as well as the necessary services connected to refuse removal, abattoirs, and markets (CPK 1927). As such, in the interwar period local governments depended heavily on the central state for financial transfers and direct provision of essential services.

Given these institutional dynamics, what kinds of powers did colonial administrators secure over urban food systems? The 1917 Nairobi Township Rules were typical in this respect (EAP 1917). They granted municipal and health officials considerable powers to license and regulate the places where food was prepared, kept, and traded. These included Indian and African "eating houses", marketplaces, bakeries, abattoirs, butcheries, fishmongers, and vegetable dealers. Licensees of eateries, for example, were required to ensure their premises were sufficiently lit and ventilated, and constructed with hard, impermeable floors. All foodstuffs were to be kept within a dedicated storeroom. Here a particular concern was to prevent people sleeping in any room in which food was cached, prepared, or sold. Anyone suffering from an infectious disease was prohibited from working on the site of a butcher, baker, or any other kind of food merchant, and food articles deemed "unfit for human consumption" could not be kept on the premises. The Rules further required the licensing of all street vendors and prohibited the hawking of any meat or fresh foods outside of public markets. Broadly speaking, the town clerk and other municipal officials were responsible for issuing licenses and managing local marketplaces, while medical officers of health and sanitary inspectors were expected to examine premises, identify nuisances, and serve legal orders.

The 1921 Public Health Ordinance generalized these kinds of rules to the territory as a whole, and was the culmination of the growing public and official enthusiasm for public health intervention that preceded the First World War. This legislation granted the state (in the form of the governor, acting upon the advice of a Central Board of Health comprising senior government officials) extensive powers to regulate: the inspection and disposal of food being imported into or exported from the colony; the construction, occupation, and use of buildings in which food was prepared or stored; standards of foodstuffs offered for sale or consumption (described in cryptic degrees of "cleanliness", "adulteration", "taint", "contamination", "soundness", and "wholesomeness"); the inspection of animals and places where animal products were prepared or stored; the quality and labelling of food receptacles; and the establishment and operation of markets (CPK 1921: 856–859). Although in principle the Ordinance allocated some public health responsibilities to local authorities, in practice these functions remained the preserve of the central Medical Department.

Public health administration only became the full responsibility of Nairobi's municipal council towards the end of the decade, following the recommendations of the Local Government Commission (CPK 1927) and passage of the Local Government (Municipalities) Ordinance in 1928. A municipal Public Health Department for Nairobi was duly established in 1929, with its own staff of medical officers and inspectors. But again, Nairobi municipality was an outlier in being appointed as a full public health authority. For the most part, public health administration remained the full or (in the case of municipal boards – second-tier local government bodies below the rank of full municipal councils) subsidized function of the state Medical Department.

While colonial public health legislation expressed an official desire for control over urban development, trade, and sanitation, in the same period the relationship between urban food and health was emerging in the form of a different problem. This was a more statistical, calculative interest in urban real incomes and the cost and standard of living (Guyer 1987b). The Kenyan Medical Department began to show concern over the health problems of urban Africans in the 1920s (Van Zwanenberg 1972). Towards the end of the decade, this dovetailed with an emerging colonial interest in African diet and nutrition – an interest originally linked to questions around how to boost the productivity of African labour (Brantley 1997). In 1933, the Department published a recommended schedule of what constituted an "adequate diet composed of local foodstuffs", based on the British Medical Association's guidelines for the minimum dietary requirements of the British working class (Van Zwanenberg 1972: 22). The problem, however, was that this diet would be unaffordable at the wage rates going in Nairobi. Indeed, it was found that food was the most costly item of expenditure for African residents. Municipal officials in Nairobi worried that the low ratio of wages to cost of living provided conditions for rapid labour turnover, malnutrition, low working productivity, endemic petty crime, and prostitution (Van Zwanenberg 1972). Despite these concerns, they had few means at their disposal to address the deficiencies of African standards of living – regulating wage rates and food prices certainly did not fall within the limited ambit of local government functions.

These points indicate how colonial officials were starting to view the problem of urban food insecurity as one of economic access and malnutrition, rather than absolute food shortage and starvation, as was the case with rural famine. This was so despite the fact that the economic depression of the 1930s was primarily a rural crisis, as real wages for Nairobi workers actually rose during the early part of the decade, largely due to falling commodity prices on export and domestic markets (Cooper 1987). These trends undoubtedly contributed to the relative lack of official urgency to deal with urban cost of living and nutrition issues. In the early 1940s however, wartime conditions and labour mobilization would reorient official attention towards urban wages and food prices. A statistical interest in measuring and regulating food prices and income as a basis of health and productivity thus emerged in parallel to a more economic interest in managing prices and distribution during times of scarcity (as discussed in the previous section). Both themes would be consolidated over the course of the Second World War.

War and urban hunger

With the outbreak of war, the Kenyan state secured wide powers of control and compulsion over food production and distribution under the 1939 Defence Regulations. A range of central institutions were formed to manage the production, supply, distribution, and pricing of key commodities. However, many of the powers claimed under the Regulations went unused during the first years of combat (Colonial Office 1941). Poor weather conditions over the course of 1942 changed the situation drastically as maize crops failed across much of the territory. By the end of the year, many British African colonies, including Kenya, found themselves in the grips of a severe food shortage. The crisis was partly due to the increased scale of demand in the territory stemming from the wartime growth of industries and labour forces, immigration, and the presence of armed forces in the region (CPK 1943). Urban populations had grown in step. Indeed, the African population of Mombasa District increased from around 40 000 in 1939 to 63 000 by the end of the war (Cooper 1987). Yet urban areas were still limited in size. In 1948, even after Kenya's wartime urban growth, township residents only constituted around 5 percent of the total population (Ogonda 1992).

The state's response to the scarcity was novel in some aspects, but familiar in others. Authorities once again prohibited the movement of key African foodstuffs between administrative districts in a bid to promote local self-sufficiency. To increase food production, special legislation was passed with invigorated economic incentives. These included offering farmers minimum returns (in the event of harvest failure) and guaranteed prices for their produce, as well as providing grants for extending cultivated areas (Lonsdale 1986). Moreover, steps were taken to decrease the rate of food consumption: workers' daily rations were reduced in quantity, and many employers retrenched parts of their labour force. Some acts were more coercive. In early 1943, for example, officials decided that food shortages in towns called for the "repatriation" of Africans to the rural reserves where, it was thought, they would have a greater chance of obtaining food. In total, around 10 000 people (including the families of workers) were sent from Nairobi to the rural areas. The approach followed was to simply deny women and children access to food rations (CPK 1943).

The higher prices available to producers led to retail price increases for key foodstuffs and consumables (Lonsdale 1986). In response, officials enacted maize price controls in July 1942. For consumers, price increases were aggravated by black marketing, which by the end of the year had become widespread in Kenyan towns, and a subject of considerable public concern (East African Standard 1942). High prices and food shortages affected those without regular employment most acutely, but all felt the effects of "rampant inflation", particularly in food and housing prices (Anderson and Throup 1985). Increased living costs and food shortages in Mombasa triggered rolling strikes, practically closing the port in October 1942 (Clayton and Savage 1974; Cooper 1987). Strikes quickly spread to other towns. This agitation, coupled with the growth of black marketing, led the authorities to form the Kenya Commodity Distribution Board in early 1943, which then implemented a "hastily conceived" rationing scheme for "starch foods" in urban areas (Grazebrook 1943). Rationing was thought to be the only means of ensuring the equitable distribution of staple foodstuffs and of keeping prices within a reasonable threshold. However, the Kenyan problem of high urban living costs relative to wages remained pressing, even being drawn into political debates in Britain (Greenridge 1943).

Unable to lower prices for fear of disincentivizing production, Kenyan authorities responded by attempting to increase the purchasing power of consumers. On one hand, this entailed charging recipients of rural famine relief at subsidized prices (O'Leary 1980). While this can be seen as a pragmatic state response to an immediate problem, the use of subsidization as an anti-scarcity measure indicates the growing extent to which people relied upon the cash purchase of food as part of their household coping strategies. On the other hand, officials responded to urban labour demands by factoring food and housing inflation into revised wage levels, and by providing a special war bonus for railway and government workers. Nutritional science was drawn into these efforts to objectively and "scientifically" determine poverty levels and wage structures (Cooper 1987). Ensuring adequate economic access to food and nutrition assumed an important role in subsequent efforts to shape a productive and stable urban working class in post-war Kenya. These efforts became the preserve of a new government Labour Department, formed in 1940 in response to the previous year's general strike in Mombasa, working in concert with the state's Medical Department.

Conclusion

In this chapter I have tried to give some historical context as to why urban governance in Africa has had limited engagement with food systems, using the specific example of Kenya. There, the establishment and consolidation of British authority coincided with periods of severe scarcity and famine. As such, for at least the first three decades of colonial rule, preventing rural famine was the administration's foremost anxiety. Municipal and central officials were concerned with food-related problems in the colony's growing towns, particularly in relation to health issues, yet in most cases administering urban public health was the responsibility of the central state. Even in the rare cases when local governments did assume public health responsibilities, as with Nairobi municipality in the late 1920s, their food-related interventions were practically limited to regulating the production and use of the built environment in order to prevent the contamination of food supplies and the spread of communicable disease. Wider aspects of the food system, including supply and distribution, lay well beyond the remit of local government, and were left largely to the incentives of private producers and traders.

When urban food insecurity emerged as a distinct problem from the late 1920s, it was seen as an issue of malnutrition and economic access, subsumed within the wider problems of how to increase standards of living, boost labour productivity, and ensure urban order. Such concerns appeared to recede in importance as officials faced the exigencies of economic depression through the 1930s. Wartime crises, however, re-emphasized the political and economic risks of urban food insecurity, but now these risks were firmly linked to labour mobilization and unrest in key towns such as Mombasa and Nairobi. Addressing the urban labour/food nexus called for interventions extending beyond the competencies of local government. These interventions encompassed matters of production, pricing, territory-wide distribution, and wage structures – the domain of a highly centralized, bureaucratized system for marketing key food supplies and managing labour that was built up to protect the colony's industries and resources for the war effort. The problem, for officials, was how to boost food production while ensuring that urban residents had access to food supplies and could afford them. Their response entailed subsidizing food purchases, as well as "scientifically" calculating wage levels. Such measures helped to establish a specific economic mode of dealing with urban food scarcity that remained a keystone of the state's marketing and food policy long after the war and colonial rule had come to an end (Maxwell 1999).

In Kenya, urban governance has engaged little with the food system due to two main trends in colonial institutional development. One encompassed the limited responsibilities and powers historically allocated to local government to deal with systemic food processes. The other was the centralization of food governance instructed by wartime conditions of the mid-twentieth century. It is against this historical background that recent interest in, and critique of, the role of urban governance and planning in promoting food security in Kenya, as well as in other post-colonial settings of sub-Saharan Africa, should be located.

References

Anderson, D. and Throup, D. (1985). Africans and agricultural production in Kenya: The myth of the war as a watershed. *Journal of African History*, 26(4), pp. 327–345.

Brantley, C. (1997). Kikuyu-Maasai nutrition and colonial science: The Orr and Gilks Study in late 1920s Kenya revisited. *The International Journal of African Historical Studies*, 30(1), pp. 49–86.

Bryceson, D.F. (1988). *Food insecurity and the social division of labour in Tanzania, 1919–1985*. PhD, University of Oxford.

Campbell, C. (2002). Juvenile delinquency in colonial Kenya, 1900–1939. *The Historical Journal*, 45(1), pp. 129–151.

Carswell, G. (2003). Food crops as cash crops: The case of colonial Kigezi, Uganda. *Journal of Agrarian Change*, 3(4), pp. 521–551.

Clayton, A. and Savage, D.C. (1974). *Government and labour in Kenya, 1895–1963*. London: Frank Cass.

Colonial Office, (1941). *Extract from summarized report on food production in the colonial dependencies during the war*. [memorandum] DO 35/848/6. Kew: The National Archives of the United Kingdom.

Colony and Protectorate of Kenya (CPK), (1921). *The official gazette of the colony and protectorate of Kenya*. Nairobi: Government Printer.

Cooper, F. (1987). *On the African waterfront: Urban disorder and the transformation of work in colonial Mombasa*. New Haven, CT and London: Yale University Press.

CPK, (1927). *Report of the local government commission*, Vol. 1. Nairobi: Government Printer.

CPK, (1929). *Report for 1928*. Nairobi: Government Printer.

CPK, (1943). *Report of the food shortage commission of inquiry*. Nairobi: Government Printer.

Davis, M. (2001). *Late Victorian Holocausts: El Niño famines and the making of the third world*. London and New York, NY: Verso.

EAP (East Africa Protectorate), (1910). *Report for 1908–09*. London: HM Stationery Office.

EAP, (1917). *The Official Gazette of the East Africa Protectorate*, XIX(553), Nairobi, pp. 833–908.

East African Standard, (1942). *Black Market Scandal*. [newspaper extract] CO 852/500/14, no. 2. Kew: The National Archives of the United Kingdom.

Gardner, L.A. (2012). *Taxing colonial Africa: The political economy of British imperialism*. Oxford: Oxford University Press.

Grazebrook, W. (1943). *Confidential report of the controller of prices and military contracts on price control in Kenya colony*. [manuscript] CO 852/500/2. Kew: The National Archives of the United Kingdom.

Greenridge, C.W.W. (1943). *Food supplies for the native population of Northern Rhodesia and Kenya*. [letter] CO 852/428/3. Kew: The National Archives of the United Kingdom.

Greenwood, A. and Topiwala, H. (2015). *Indian doctors in Kenya, 1895–1940: The forgotten history*. Basingstoke: Palgrave Macmillan.

Guyer, J.I., ed., (1987a). *Feeding African cities: Studies in regional social history*. Bloomington, IN: Indiana University Press.

Guyer, J.I. (1987b). Introduction. In: J.I. Guyer, ed., *Feeding African cities: Studies in regional social history*. Bloomington, IN: Indiana University Press, pp. 1–54.

Holm, A. (1929). *Report on the locust position*. [letter] CO 533/384/2, n. 24. Kew: The National Archives of the United Kingdom.

Home, R. (2012). Colonial township laws and urban governance in Kenya. *Journal of African Law*, 56(2), pp. 175–193.

Iliffe, J. (1990). *Famine in Zimbabwe, 1890–1960*. Gweru: Mambo Press.

Iliffe, J. (2007). *Africans: The history of a continent*, 2nd ed. Cambridge: Cambridge University Press.

Lonsdale, J. (1986). The depression and the second world war in the transformation of Kenya. In: D. Killingray and R. Rathbone, eds., *Africa and the second world war*, 1st ed. Basingstoke: Palgrave Macmillan, pp. 97–142.

Maxwell, D. (1999). The political economy of urban food security in sub-Saharan Africa. *World Development*, 27(11), pp. 1939–1953.

Njoh, A. (2009). Urban planning as a tool of power and social control in colonial Africa. *Planning Perspectives*, 24(3), pp. 301–317.

Ogonda, R.T. (1992). Transport and communications in the colonial economy. In: W.R. Ochieng and R.M. Maxon, eds., *An economic history of Kenya*, 1st ed. Nairobi: East African Educational Publishers, pp. 129–145.

O'Leary, M. (1980). Responses to drought in Kitui district, Kenya. *Disasters*, 4(3), pp. 315–327.

Porter, G., Lyon, F. and Potts, D. (2007). Market institutions and urban food supply in west and Southern Africa: A review. *Progress in Development Studies*, 7(2), pp. 115–134.

Robertson, C.C. (1997). *Trouble showed the way: Women, men, and trade in the Nairobi Area, 1890–1990*. Bloomington, IN: Indiana University Press.

Silva, C.N., ed., (2015). *Urban planning in sub-Saharan Africa: Colonial and post-colonial cultures*. New York, NY and London: Routledge.

Van Zwanenberg, R. (1972). *History and theory of urban poverty in Nairobi: The problem of slum development*. Discussion Paper no. 139. Nairobi: Institute for Development Studies, University of Nairobi.

Van Zwanenberg, R. and King, A. (1975). *An economic history of Kenya and Uganda, 1800–1970*. Basingstoke and London: Macmillan.

Vernon, J. (2007). *Hunger: A modern history*. Cambridge, MA, and London: Belknap Press.

6 Current urban food governance and planning in Africa

Warren Smit

Introduction

This chapter provides an overview of the governance of urban food systems in Africa, and sets the scene for the following five chapters, which look at particular aspects of the governance of urban food systems in each of the case study cities. Understanding the existing governance of urban food systems is an essential precondition for attempting to improve governance processes so as to contribute to enhanced urban food security. A particularly important aspect of governance with regards to urban food systems is urban planning. Problems with governance and urban planning processes are arguably a key contributor to the high levels of food insecurity in African cities. As Porter *et al.* (2007: 116) suggest, there are complex processes and rules shaping food systems in Africa, and "we need to know more about how these formal and informal regulatory systems operate if we are to improve access to markets and thus enhance urban food supplies and also secure income and livelihoods." The New Urban Agenda, adopted by the United Nations (UN) in 2016, similarly recognises the need for strengthening governance, strengthening food system planning and "improving food security and nutrition" in urban areas (UN 2016: 11).

In this chapter, drawing on reviews of the food and governance literature on Africa and on fieldwork in Kenya, Zambia, and Zimbabwe, I provide a broad overview of the impact of governance and planning processes on urban food systems in Africa. First of all, I discuss the concept of "governance". I then discuss the linkages between urban governance/planning and urban food systems in Africa, with a focus on four key aspects of governance of urban food systems: the governance of the production of food in and around cities, with a focus on urban and peri-urban agriculture; the governance of food distribution; the governance of formal and informal food retail; and the governance of food safety, both in terms of the safety of the food itself and the safety of food production and preparation processes. In general, governance and planning processes have an ambiguous impact on urban food systems and urban food security, and governance and planning processes in relation to food need to be made more effective (although it should be noted that urban governance dynamics and urban food systems vary considerably from country to country, and also within countries). Finally, I identify some potential ways we can approach improving the governance and planning of urban food systems in Africa.

Urban governance of food systems

Governance can be defined in various ways. Many definitions, for example those of the World Bank, are closely linked with normative notions of "good governance", and arguably focus too much on what those organisations think should happen rather than on what actually happens in reality (Doornbos 2001; Pearce 2003). Many scholars have found it more useful to use the concept of governance in a more analytical way, for example, as "the range of political, organisational, and administrative processes through which stakeholders (including citizens and interest groups) articulate their interests, exercise their legal rights, take decisions, meet their obligations, and mediate their differences" (Bakker *et al.* 2008: 1894).

One of the key characteristics of using a governance lens is the recognition that there is a wide range of actors involved in governance, such as various government organisations, political parties, traditional leaders, civil society organisations – for example, non-governmental organisations (NGOs) and community groups – and the private sector (Devas 2001). Using a governance lens is essentially about understanding these actors and the relationships between them. The distribution of power can be conceived of as being dispersed amongst these different actors and "exercised from innumerable points" (Foucault 1976/1998: 94) by individuals such as "planners, policy analysts and researchers, and politicians" (Richardson 1996: n.p.). Power is not evenly distributed, however, as there are particular "centres" or "nodes" with concentrations of power (Lindell 2008). These governance nodes impact on cities and towns through a range of "formal" and "informal" decision-making and regulatory processes. Formality/informality should be regarded as a continuum rather than a dichotomy, with different levels of formality and informality often overlapping, but I use the terms to refer to the respective halves of the continuum. As Devas (2004) notes, informal governance processes are, in practice, often more important than formal governance processes.

The actors involved in urban governance have very different interests and agendas, and there are few processes for reaching consensus or resolving competing interests. The net result is often chaotic. As Lindell (2008: 1896) notes of the governance of market places in Maputo, Mozambique, "governance appears to lack any semblance of coherence and to be more fragmented, disjointed and split by deep antagonisms. The key actors continuously challenge each other's legitimacy to 'govern.'" The governance and planning of urban food systems is particularly complex as urban food systems are generally not thought of as an explicit object of governance, and the impact of governance and planning systems on urban food systems is usually unintentional (Pothukuchi and Kaufman 2000). The governance of food systems is also complex and challenging because of "the interdependencies of actors and their activities, as well as their different framings of the problem of food insecurity" (Pereira and Drimie 2016: 25–26). It is only in recent decades, mainly in the global North, that actors in urban governance have begun to explicitly think about urban food systems and how to promote urban food security.

Urban governance actors can impact on urban food systems, and thus on urban food security, in a variety of ways; for example, through the governance of: the production of food in and around cities; the functioning of urban and peri-urban transport systems, which impact on food distribution; formal and informal food retail; and food safety. These four areas of governance are discussed in the following four sections of the chapter.

Governance of urban and peri-urban agriculture in Africa

Urban governance can have a big impact on the extent and nature of agriculture in and around cities. The importance of urban agriculture varies considerably between cities and countries in Africa, and is often quite limited (Crush *et al.* 2011; Zezza and Tasciotti 2010). Part of the reason for many African cities having fairly low levels of urban agriculture is that most African national and local governments have been intolerant of urban agriculture, seeing it as incompatible with their "modernist" visions of what cities should look like (Simatele and Binns 2008). The attitude of local government in Lusaka is typical of many local governments in Africa; at the time, they viewed urban agriculture as a rural activity "whose practice within the city boundaries is inappropriate and detracts from the modern image of the city" (Simatele and Binns 2008: 2).

In cities where urban agriculture is tolerated, organisations of urban farmers can play an important role in promoting and supporting urban agriculture. In Dar es Salaam, for example, the activities of urban farmers' associations include "joint production on communally held property and serving as a loan or finance agency for their members" (Schmidt *et al.* 2015).

Some cities, such as Cape Town and Ndola, have developed urban agriculture policies which attempt to support low-income households' participation in urban agriculture. Local government support is potentially important in enabling low-income households to be able to engage in urban agriculture, through the provision of "inputs, extension services, credit/financial access, production and marketing infrastructure, and knowledge" (Frayne *et al.* 2014: 187).

Controls over land use in peri-urban areas are often laxer than is the case in central urban areas. Many functional urban areas in Africa include large expanses of peri-urban agricultural land, and African cities are often characterised by "unregulated peri-urban land development" that has a negative impact on peri-urban agriculture (Kombe 2005: 113). The ability of local government to control such sprawl is often limited by peri-urban areas being in customary ownership and under traditional authorities (e.g. Gough 1999). The net result is the gradual displacement of peri-urban agriculture. For example, in Ghana, "as urbanisation increases, farmers are being pushed unto less favourable lands, farther villages or restricted to unauthorised public spaces in order to continue production. The absence of urban green belts reduces farming to flood plains and along public drains" (Kuusaana and Eledi 2015: 462).

Governance of food distribution

The distribution of food is essentially about "how food for consumption is physically moved to be available, in what form, when and to whom" (Ericksen 2008: 239). The key determinants of food distribution include "transportation infrastructure, trade regulations, government transfer programs, and storage requirements" (Ericksen 2008: 238).

At the urban scale, the key determinant of food distribution is the urban (and peri-urban) transport system, the key elements of which (provision and maintenance of roads, managing traffic, ensuring roadworthiness of vehicles) are the responsibility of local (or other levels of) government. In general, urban transport infrastructure in African cities is inadequate. A study of 14 cities in Africa found that "to a greater or lesser extent in each of the cities, the networks of paved roads and associated traffic control facilities are deficient" (Kumar and Barrett 2008: 37). Systems and procedures for ensuring vehicle safety are also largely inadequate (Kumar and Barrett 2008). Although difficult to quantify, it is clear that "in sub-Saharan Africa, limited infrastructure and transport service has occasionally disrupted food production and circulation" (Pirie 1993: 12). There have been a few documented instances where inadequate availability of functioning vehicles and inadequate maintenance of roads have significantly hindered the distribution of food (e.g. Good 1988). The limitations of transport infrastructure in and around African cities can also increase the cost transportation and thus also increase the cost of urban food (Gollin and Rogerson 2010).

Governance of food retail

Most of the food retail sector in most African cities can be classified as being at the "informal" end of the formality/informality continuum. There are usually a wide variety of different types of informal food retail outlets, such as large traditional market places (which might have thousands of traders), various types of informal shops and kiosks, and street food vendors (see Chapter 7). Urban governance can impact on informal food traders in various ways, for example, through processes for assigning land uses, allocating trading space and through the provision of infrastructure and services, such as water, electricity and refuse removal.

Market places are a particularly important element of urban food systems in Africa and are an important site of urban governance. The governance of market places can impact on the accessibility, affordability and quality of food. Market associations generally play an important role in managing market places in Africa, particularly in West Africa (King 2006; Porter *et al.* 2007). Market associations "control the selling space and can therefore exclude others and have wider effects on the vegetable production and marketing system" (Lyon 2003:

20). In Maputo, for example, "the market committees provide infrastructure (water, toilets, etc.), maintenance and security services, and organise cleaning in their respective markets. . . . The committees also act as the principal regulators in the markets" (Lindell 2008: 1889). Market associations can also influence the decisions of local governments; for example, the Market Traders Association in Kumasi, Ghana, successfully opposed plans of the local government to increase market fees by 300% (King 2006). Local governments also usually play a role in managing market places, partially because trader fees can be a significant source of local government revenue (King 2006). As most local governments lack adequate power to plan, regulate and provide infrastructure and services, though, this role is often limited to collecting fees (Meagher 2011).

Street food vendors play an important role in providing a variety of food for the urban poor (Dixon *et al.* 2007; Steyn *et al.* 2013), but, whereas traders in market places have some protection, as a result of numbers and some form of official recognition, street traders are much more vulnerable. There are frequently waves of evictions of street traders, for example, when there is a change of political power or when a major event is about to take place (Hansen 2004; King 2006; Potts 2007; Setsabi 2006). A number of scholars have seen this frequent intolerance of informal traders as similar to the widespread prevention of urban agriculture, in that it is driven by modernist conceptualisations of urban planning as being about correcting what politicians and planners perceive as the unruly and chaotic landscapes of the informal sector (Kamete 2013; Rogerson 2016).

Over time, the nature of food retail in African cities has been changing, with increased growth of "formal" retail outlets. In the past two decades, there has been rapid growth in supermarkets in Africa (Crush and Frayne 2011; Weatherspoon and Reardon 2003; also see Chapter 2). The implications of this transformation for urban food security are not well understood, but what evidence exists suggests that the shift from local production of food and a largely informal retail sector to formal supermarkets with international supply chains may result in decreased levels of food security (Minten 2008; Riley and Legwegoh 2014).

Currently, the roll-out of supermarkets is being driven by the private sector; Crush and Frayne (2011: 806) note that there is an "absence of regulatory controls on supermarket expansion in urban markets" in Africa. However, local government potentially has an important role to play in deciding where supermarkets are located, how big they are, how they are designed (for example, whether they are accessible to pedestrians) and whether they offer surrounding space for informal traders.

Governance of food safety

The main determinants of food safety, and the safety of those involved in producing and processing food, are "the procedures and standards and regulations (or lack of) for food production, processing, and packaging" (Ericksen 2008: 240). Regulating food safety is therefore an important component of the governance of urban food systems (also see Chapter 12). Many food safety regulations need to be enforced at a national level, for example, the formulation and enforcement of standards on the composition of imported and locally manufactured food. In this section, though, I focus on a particular aspect of the governance of food safety, namely the governance of the safety of food sold by informal traders.

Although street food can be healthy and hygienic (von Holy and Makhoane 2006), and forms a key part of the diet of many African urban residents (Steyn *et al.* 2013), the safety of street foods is often seen as an area of concern by local policy makers, as street foods "often do not meet proper hygiene standards, in large part because of weak regulatory systems, inadequate food safety laws, lack of financial resources to invest in safer equipment, and lack of education for food-handlers" (CSPI 2005: 25). Inadequate infrastructure is also a key problem; studies from a number of African countries (e.g. Cambaza dos Muchangos *et al.* 2015; Muyanja *et al.* 2011) have confirmed that street food can have a high risk of contamination where there is a lack of appropriate infrastructure. Typical issues include inadequate access to water and sanitation, inadequate refuse removal and exposure to flies, which can all result in contamination of food.

Some, but not all, local governments are involved in enforcing standards of hygiene amongst informal traders (Muyanja *et al.* 2011). Even where local government does attempt to enforce health standards, this enforcement is often only partial. For example, in Abeokuta, Nigeria, food vendors are required to obtain an annual certificate from health authorities, but a survey found that only 31% of vendors had these (Omemu and Aderoju 2008).

Most recommendations for ensuring food safety focus on hygienic practices by food handlers (e.g. WHO 2006), and it is important for stakeholders in urban governance to raise awareness of this and ensure implementation (for example, through inspection and penalties), but it is arguably even more important to ensure that the necessary infrastructure and facilities are available. A study of street food vendors in Tshwane, South Africa, found that vendors generally followed "good basic hygiene practices" but that the overall environment in which they cooked and sold food resulted in the contamination of food (Oguttu *et al.* 2015: 204). Providing potable water, sanitation and adequate protection from the elements for markets and informal traders is therefore important for helping to ensure food safety.

Improving governance and planning of urban food systems

It is only in recent decades, mainly in the global North, that actors in urban governance have begun to explicitly think about urban food systems and how to promote urban food security. There have been numerous attempts at the collaborative governance of urban food systems in the global North, such as setting up Food Policy Councils in an attempt to coordinate various governance actors in ensuring that urban food systems promote food security (MacRae and Donahue 2013) and implementing urban food strategies, generally focused on issues of food quality, health and sustainability (Sonnino 2016). Urban food strategies in the global South are rare, but where they exist (for example, in Belo Horizonte, Brazil) they have a focus on ensuring affordable and healthy food for the poor. In the 1990s, the local government of Belo Horizonte launched a food security programme, the three main areas of which were: to prevent malnutrition by assisting poor households and individuals at risk; working with the private sector to ensure increased accessibility, affordability and quality of staple foods and fruit and vegetables; and increasing food production and supply (Rocha and Lessa 2009). There have been some tentative steps towards developing urban food system strategies in South Africa (Haysom 2015), but, in general, the governance of urban food systems in African cities still happens in an uncoordinated and unintegrated way, and urban planning is often used as a tool for clamping down on informality rather than promoting livelihoods and food security.

The New Urban Agenda has recognised the growing problem of urban food security and has called for "the integration of food security and the nutritional needs of urban residents, particularly the urban poor, in urban and territorial planning, in order to end hunger and malnutrition" (UN 2016: 17). In order to begin to do this in Africa, we need to better understand existing urban governance processes and the competing interests of urban governance actors in order to be able to collaboratively design interventions to improve urban food security. In particular, we need to know more about the roles that local governments in Africa do, and can, play. Local governments potentially have an important role to play in promoting urban food security, as in Belo Horizonte. Coordinating the roles of different actors involved in governing urban food systems will be important, perhaps necessitating something like the Food Policy Councils found in North America, but the frequently antagonistic relationships between governance actors in many African cities will make this challenging. An essential precondition for more effective governance of urban food systems is to bring different stakeholders together to engage with each other. Such processes need to explicitly recognise power dynamics and competing interests (Pereira and Drimie 2016). There are a number of examples of collaborative urban governance in Africa that are worth learning from, such as the example of the Kisumu Action Team in Kisumu, Kenya, which brought together a range of stakeholders (including local government, civil society and informal traders' organisations) to develop an inter-sectoral strategy for Kisumu that included a number of planned interventions relating to food, such as the upgrading of market places and promotion of local food production (Onyango and Obera 2015). There is also a need to change urban planning processes and procedures to make them more pro-poor and more supportive of the promotion of urban food security.

Conclusion

This chapter highlights the wide range of urban governance actors with very different agendas that impact on urban food systems in Africa. As a result, the governance of urban food systems is characterised by fragmentation and a lack of coordination between governance actors. This means that key problems, such as inadequate infrastructure in market places, are often not adequately addressed, and the problems persist and grow over time. There is an urgent need to bring together urban governance actors to develop and implement coherent strategies for improving urban food security in a sustainable way. Examples such as the Kisumu Action Team show how it possible to bring together urban governance actors to develop holistic strategies that attempt to have a positive impact on urban food security and other urban challenges. It is only through collaborative governance processes that we can begin to develop well-functioning urban food systems in Africa.

References

Bakker, K., Kooy, M., Shofiani, N.E. and Martijn, E.J. (2008). Governance failure: Rethinking the institutional dimensions of urban water supply to poor households. *World Development*, 36(10), pp. 1891–1915.

Cambaza dos Muchangos, A.B., Roesel, K., McCrindle, C., Matusse, H., Hendrickx, S., Makita, K. and Grace, D. (2015). Informal markets in Mozambique risky for local chicken. In: K. Roesel and D. Grace, eds., *Food safety and informal markets: Animal products in sub-Saharan Africa*. Abingdon: Routledge, pp. 197–200.

Crush, J. and Frayne, B. (2011). Supermarket expansion and the informal food economy in Southern African cities: Implications for urban food security. *Journal of Southern African Studies*, 37(4), pp. 781–807.

Crush, J., Hovorka, A. and Tevera, D. (2011). Food security in Southern African cities: The place of urban agriculture. *Progress in Development Studies*, 11(4), pp. 285–305.

CSPI (Center for Science in the Public Interest), (2005). *Food safety around the World*. Washington, DC: CSPI.

Devas, N. (2001). Does city governance matter for the urban poor? *International Planning Studies*, 6(4), pp. 393–408.

Devas, N. (2004). Urban poverty and governance in an era of globalization, decentralization and democratization. In: N. Devas, ed., *Urban governance, voice and poverty in the developing world*. London: Earthscan, pp. 15–36.

Dixon, J., Omwega, A., Friel, S., Burns, C., Donati, K. and Carlisle, R. (2007). The health equity dimensions of urban food systems. *Journal of Urban Health*, 84(1), pp. 118–129.

Doornbos, M. (2001). 'Good governance': The rise and decline of a policy metaphor? *The Journal of Development Studies*, 37(6), pp. 93–108.

Ericksen, P.J. (2008). Conceptualizing food systems for global environmental change research. *Global Environmental Change*, 18(1), pp. 234–245.

Foucault, M. (1998). *The will to knowledge: The history of sexuality*, Vol. 1 (R. Hurley, Trans.). London: Penguin (Original work published 1976).

Frayne, B., McCordic, C. and Shilomboleni, H. (2014). Growing out of poverty: Does urban agriculture contribute to household food security in Southern African cities? *Urban Forum*, 25(2), pp. 177–189.

Gollin, D. and Rogerson, R. (2010). Agriculture, roads and economic development in Uganda. *National Bureau of Economic Research, Working Paper 15863*. Cambridge, MA: NBER.

Good, K. (1988). Weak state and backward agriculture in Zambia: A case study and its Implications. In: K. Chazan and T.M. Shaw, eds., *Coping with Africa's food crisis*. Boulder, CO: Lynne Rienner, pp. 185–209.

Gough, K. (1999). The changing nature of urban governance in peri-urban Accra, Ghana. *Third World Planning Review*, 21(4), pp. 393–410.

Hansen, K.T. (2004). Who rules the streets? The politics of vending space in Lusaka. In: K.T. Hansen and M. Vaa, eds., *Reconsidering informality: Perspectives from Urban Africa*. Uppsala: Nordic Africa Institute, pp. 62–80.

Haysom, G. (2015). Food and the city: Urban scale food system governance. *Urban Forum*, 26(3), pp. 263–281.

Kamete, A.Y. (2013). Missing the point? Urban planning and the normalisation of 'pathological' spaces in southern Africa. *Transactions Institute of British Geographers*, 38, pp. 639–651.

King, R. (2006). Fulcrum of the urban economy: Governance and street livelihoods in Kumasi, Ghana. In: A. Brown, ed., *Contested space: Street trading, public space, and livelihoods in developing cities*. Rugby: ITDG Publishing, pp. 100–118.

Kombe, W.J. (2005). Land use dynamics in peri-urban areas and their implications on the urban growth and form: The case of Dar es Salaam, Tanzania. *Habitat International*, 29(1), pp. 113–135.

Kumar, A. and Barrett, F. (2008). *Stuck in traffic: Urban transport in Africa*. Africa Infrastructure Country Diagnostic report. Washington, DC: The World Bank.

Kuusaana, E.D. and Eledi, J.A. (2015). As the city grows, where do the farmers go? Understanding Peri-urbanisation and food systems in Ghana—Evidence from the Tamale Metropolis. *Urban Forum*, 26(4), pp. 443–465.

Lindell, I. (2008). The multiple sites of urban governance: Insights from an African City. *Urban Studies*, 45(9), pp. 1879–1901.

Lyon, F. (2003). Trader associations and urban food systems in Ghana: Institutionalist approaches to understanding urban collective action. *International Journal of Urban and Regional Research*, 27(1), pp. 11–23.

MacRae, R. and Donahue, K. (2013). *Municipal food policy entrepreneurs: A preliminary analysis of how Canadian cities and regional districts are involved in food system change*. Toronto: Toronto Food Policy Council & Canadian Agri-Food Policy Institute.

Meagher, K. (2011). The tangled web of associational life: Urban governance and the politics of popular livelihoods in Nigeria. *Urban Forum*, 21(3), pp. 299–313.

Minten, B. (2008). The food retail revolution in poor countries: Is it coming or is it over? *Economic Development and Cultural Change*, 56(4), pp. 767–789.

Muyanja, C., Nayiga, L., Brenda, N. and Nasinyama, G. (2011). Practices, knowledge and risk factors of street food vendors in Uganda. *Food Control*, 22(10), pp. 1551–1558.

Oguttu, J., Roesel, K., McCrindle, C., Hendrickx, S., Makita, K. and Grace, D. (2015). Arrive alive in South Africa: Chicken meat the least to worry about. In: K. Roesel and D. Grace, eds., *Food safety and informal markets: Animal products in sub-Saharan Africa*. Abingdon: Routledge, pp. 202–220.

Omemu, A. and Aderoju, S. (2008). Food safety knowledge and practices of street food vendors in the City of Abeokuta, Nigeria. *Food Control*, 19(4), pp. 396–402.

Onyango, G.M. and Obera, O.B. (2015). Tracing Kisumu's path in the co-production of knowledge for urban development. In: M. Polk, ed., *Co-producing knowledge for sustainable cities: Joining forces for change*. Abingdon: Routledge, pp. 73–97.

Pearce, K.C. (2003). Democracy and development as contested terrain: The discourse of the 'good governance' agenda in Africa. *Review of Communication*, 3(1), pp. 28–30.

Pereira, L. and Drimie, S. (2016). Governance arrangements for the future food system: Addressing complexity in South Africa. *Environment: Science and Policy for Sustainable Development*, 58(4), pp. 18–31.

Pirie, G.H. (1993). Transport, food insecurity, and food aid in sub-Saharan Africa. *Journal of Transport Geography*, 1(1), pp. 12–19.

Porter, G., Lyon, F. and Potts, D. (2007). Market institutions and urban food supply in West and Southern Africa: A review. *Progress in Development Studies*, 7(2), pp. 115–134.

Pothukuchi, K. and Kaufman, J.L. (2000). The food system: A stranger to the planning field. *Journal of the American Planning Association*, 66(2), pp. 113–124.

Potts, D. (2007). City life in Zimbabwe at a time of fear and loathing: Urban planning, urban poverty and Operation Murambatsvina. In: G. Myers and M. Murray, eds., *Cities in contemporary Africa*. New York, NY: Palgrave Macmillan, pp. 265–288.

Richardson, T. (1996). Foucauldian discourse: Power and truth in urban and regional policy making. *European Planning Studies*, 4(3), pp. 279–292.

Riley, L. and Legwegoh, A. (2014). Comparative urban food geographies in Blantyre and Gaborone. *African Geographical Review*, 33(1), pp. 52–66.

Rocha, C. and Lessa, I. (2009). Urban governance for food security: The alternative food system in Belo Horizonte, Brazil. *International Planning Studies*, 14(4), pp. 389–400.

Rogerson, C.M. (2016). Suth Africa's informal economy: Reframing debates in national policy. *The Journal of the Local Economy Policy Unit*, 31(1–2), pp. 172–186.

Schmidt, S., Magigi, W. and Godfrey, B. (2015). The organization of urban agriculture: Farmer associations and urbanisation in Tanzania. *Cities*, 42(Part B), pp. 153–159.

Setsabi, S. (2006). Contest and conflict: Governance and street livelihoods in Maseru, Lesotho. In: A. Brown, ed., *Contested space: Street trading, public space, and livelihoods in developing cities*. Rugby: ITDG Publishing, pp. 131–148.

Simatele, D.M. and Binns, T. (2008). Motivation and marginalization in African urban agriculture: The case of Lusaka, Zambia. *Urban Forum*, 19(1), pp. 1–21.

Sonnino, R. (2016). The new geography of food security: Exploring the potential of urban food strategies. *The Geographical Journal*, 182(2), pp. 190–200.

Steyn, N., Mchiza, Z., Hill, J., Davids, Y., Venter, I., Hinrichsen, E., Opperman, M., Rumbelow, J. and Jacobs, P. (2013). Nutritional contribution of street foods to the diet of people in developing countries: A systematic review. *Public Health Nutrition*, 17(6), pp. 1363–1374.

UN (United Nations), (2016). *Draft outcome document of the United Nations Conference on Housing and Sustainable Urban Development (Habitat III)*. Available at: https://www2.habitat3.org/bitcache/99d99fbd0824de50214e99f864459d8081a9be00?vid=591155&disposition=inline&op=view [Accessed 27 Feb. 2018].

von Holy, A. and Makhoane, F.M. (2006). Improving street food vending in South Africa: Achievements and lessons learned. *International Journal of Food Microbiology*, 111, pp. 89–92.

Weatherspoon, D.D. and Reardon, T. (2003). The rise of supermarkets in Africa: Implications for agrifood systems and the rural poor. *Development Policy Review*, 21(3), pp. 333–355.

WHO (World Health Organization), (2006). *Five keys to safer food manual*. Geneva: WHO.

Zezza, A. and Tasciotti, L. (2010). Urban agriculture, poverty, and food security: Empirical evidence from a sample of developing countries. *Food Policy*, 35(4), pp. 265–273.

7 Contributing yet excluded?

Informal food retail in African cities

Caroline Skinner

Introduction

Despite greater supermarket penetration in many urban areas, informal food retailers – street and market traders, 'spaza' or small-scale shops – remain ubiquitous across the African continent. In urban food security debates, increasing attention is being paid to the whole urban food system, rather than simply to production (see Chapter 3), and as a result the role of the informal food sector is getting greater attention (Crush and Frayne 2011). Historically, public health research has focused on establishing the role of what they describe as 'street foods' in energy and protein intake. In contrast, the rich literature on the informal sector/economy[1] only cursorily mentions food and seldom considers urban food security implications. Bringing together disparate literatures, this chapter interrogates the informal sector–food security nexus. Three dimensions are highlighted. First, given that a key element of food accessibility is household income, labour force survey statistics are analysed to outline the extent of employment in the informal sector in general, and informal retail and informal food retail in particular. Second, what is currently known about informal retailers' role in food intake and household sourcing strategies is reviewed. Finally, the impact of supermarket penetration on food security on one hand, and informal retailers on the other, are interrogated. Together this makes the case that informal retailers are making a critical contribution to urban food security. This sets the scene for the analysis of policy and regulatory trends for informal retailers which is shown to be largely exclusionary especially for those operating in public space. Analysis of the underlying factors behind these trends are offered.

The informal sector as a source of employment

Regional informal sector estimates demonstrate that for many regions in the global South, the majority of those who work outside of agriculture[2] work in the informal sector (Vanek *et al.* 2014: 8–11). In sub-Saharan Africa, for example, 53% of non-agricultural employment is in the informal sector. Table 7.1 reflects available statistics for countries in sub-Saharan Africa showing

Table 7.1 Sub-Saharan Africa: informal sector employment as percentage of total non-agricultural employment, by sex (latest year available)

Survey year	*Country*	*Women*	*Men*	*Total*
2004	Mali	80	63	71
2008	Cote d'Ivoire	83	61	70
2008	Zambia	70	61	65
2010	Uganda	62	58	60
2005	Madagascar	64	41	52
2005/2006	Tanzania	50	53	52
2010	Liberia	65	33	50
2008	Lesotho	48	50	49
2004	Ethiopia*	48	36	41
2004	Zimbabwe	53	31	40
2010	South Africa	17	19	18

* urban areas
(*International Labour Organisation 2013*: 10, 13, 14)

significant diversity with the informal sector contribution at 70% and 71% of total non-agricultural employment in Mali and Cote d'Ivoire and only 18% in South Africa.

While the informal sector in many regions is a disproportionate source of employment for men, in sub-Saharan Africa 59% of women are in the informal sector, compared to 49% of men. The country statistics show that of the 11 countries for which there is data, only in Tanzania, Lesotho and South Africa is the informal sector a greater source of employment for men than for women. These data suggest that the informal sector in general is playing a significant role in employment generation and thus income needed to access food. Further, income in the hands of women is more likely to be spent on household needs including food (see for example Duflo 2003) and in the sub-Saharan region the informal sector is a bigger source of employment for women.

Available industrial segmentation data suggest that trade is an important source of employment in the informal sector in the global South in general and disproportionately so in sub-Saharan Africa where it constitutes 43% of total informal employment.[3] These high proportions are confirmed in available country data (ILO 2013, Annex 2). These data draw from standard labour force surveys, which often include a question to informal retailers on whether they sell food or non-food products. Wills (2009) analysed South Africa data and found that almost two-thirds of informal traders surveyed sold food.

Given the paucity of national data on informal food retail, smaller scale survey work is important. Tinker's (1997) seminal work on street foods in the mid-1990s drew on research from numerous cities including those in Egypt, Nigeria and Senegal. The surveys found that as much as 10% of the local population was employed in the informal food sector and 20% of household expenditure went towards these foods. More recent analyses of street food vendors in Africa confirm that this is a key source of employment, particularly for women, and is a critical source of household income and well-being (Chicho-Matenge and Nakisani 2013; Otoo *et al.* 2012). Albeit dated, the Levin *et al.* (1999) study unpacked the implications of money in the hands of women. In their nearly 600 household interviews conducted in Accra in Ghana, women were found to earn lower incomes but allocate more of their budget to basic goods for themselves and their children. A resounding theme in this literature is the difficult conditions under which food traders operate, often without water, toilets, shelter and electricity.

Informal retailers' role in food intake and household sourcing strategies

Existing evidence suggests food sourced informally makes a significant contribution to energy and protein intake. Steyn *et al.* (2013) review 23 studies, the majority of which were conducted in Africa (Kenya, Nigeria, Ghana, Benin, Mali, South Africa, Uganda and Burkina Faso). They found the daily energy intake from what they call 'street foods' in adults ranged from 13% to 50%, and in children from 13% to 40%. Similarly, they calculated that street foods contribute significantly to the daily intake of protein, often 50% of the recommended daily allowance (Steyn *et al.* 2013: 1). From a public health perspective, they conclude that street foods "should be encouraged"(Steyn *et al.* 2013: 1).

The household food security surveys conducted first by the African Food Security Network (AFSUN), and then more recently by the Consuming Urban Poverty (CUP) project, provide insights into the role of informal retailers in household food sourcing strategies. AFSUN surveyed low-income neighbourhoods in 11 Southern African cities. Seventy percent of 6453 households surveyed reported sourcing food from informal outlets (Crush and Frayne 2011: 799). They also established that the more food insecure a household was, the more likely it was to source food from informal retailers. There was variation by city explained as follows:

> The informal food economy proved to be particularly important in cities such as Lusaka, Harare, Blantyre and Maputo (where over 95% of poor households normally obtain food from informal sources at least once a week). Their importance varied considerably in the three South African cities (from a high of 85% in Johannesburg to a low of only 42% in Msunduzi). In Windhoek, around three quarters of households normally sourced informal food but only a half did so in Maseru and Manzini. Households in Gaborone were the least reliant on the informal food economy (at only 29%).
>
> (Crush and Frayne 2011: 789)[4]

They also note in the larger cities, where more than one area was sampled, that the differences were not as significant as expected. In Johannesburg, for example, over 80% of surveyed households sourced food from the informal sector in each of the areas of Alexandra, the Inner City and Orange Farm. Inter-city variability and the absence of intra-city variation is striking and cannot easily be explained without more research on the size and nature of the informal food sector in each city.

This research also provides critical insight into the use of supermarkets. Seventy-nine percent of interviewees reported purchasing from supermarket. Survey evidence on the frequency of purchase by outlet is presented in Table 7.2. This shows a pattern of frequent purchases in the informal sector – 32% of households in the surveyed areas patronised the informal food sector almost every day and 59% did so at least once a week. In contrast, a high proportion of respondents reported that they shopped at supermarkets monthly. This may suggest a pattern of bulk buying of staples at supermarkets and a reliance on the informal food sector for other foodstuffs (Crush and Frayne 2011: 801).

The CUP project adds further nuance to the AFSUN findings. Table 7.3 shows the distribution of household food sources by types of outlet in the three survey areas.

These data demonstrate how households make extensive use of both the formal and informal outlets in their food sourcing. Even in contexts like Kisumu where supermarket usage is high, there is continued use of home-based outlets

Table 7.2 Patronage by outlet type (% of households interviewed in the AFSUN survey)

	Informal outlet	*Supermarket*
In the previous week	60%	54%
Frequently (at least five days a week)	32%	5%
At least once a week	27%	19%
At least once a month	7%	51%
At least once every six months	1%	2%
Occasionally (less than once a year)	0.5%	1%
Never	31%	22%

(*Redrawn from Crush and Frayne 2011*: 800, 802)

Table 7.3 Percentage[5] distribution of household food sources by type of outlet in Epworth, Kisumu and Kitwe, 2016/17[6]

	Informal outlets				*Formal outlets*				
	Home-based operator	Local market	Street vendor	Kiosk	Whole-saler	Super-market	Mini-market	Butchery	N*=
Epworth, Zimbabwe	79%	22%	48%	0%	52%	33%	28%	58%	486
Kisumu, Kenya	81%	71%	20%	0%	12%	75%	27%	69%	841
Kitwe, Zambia	34%	88%	48%	51%	14%	31%	3%	8%	882

N* = the total number of households interviewed
(*CUP Household Food Security Surveys*)[7]

and local markets. What these data also suggest is that usage of outlet types varies considerably. For example, while usage of local markets is widespread in Kismu and Kitwe, it is very low in Epworth. In the formal sector, while wholesalers in Epworth and supermarkets in Kisumu are a predominant food source, in Kitwe informal outlets prevail, with comparatively small numbers of households sourcing their food from any formal retailer. In line with AFSUN findings, these data found that all food-insecure households sourced food from informal outlets. Together these data show significant variation between and within cities – a reminder that the role played by the informal food sector is likely to be context-specific.

Supermarket expansion, food security and informal retailers

While the nature and pace of supermarket penetration across the African continent is a source of debate (Reardon *et al.* 2004; Traill 2006; Humphrey 2007), few would contest that supermarkets are reconfiguring the nature of food access. Skinner and Haysom (2016: 9) for example note:

> In 2002 Shoprite had 77 retail stores in 13 African countries (excluding South Africa), but by 2012 it had expanded to 168 stores in 18 countries; by August 2013 it had expanded at the rate of one new store every two weeks to reach 193 outlets . . . by 2013 there were 3741 stores in sub-Saharan Africa.

Those who support the expansion of supermarkets argue that their greater purchasing power and economies of scale will benefit the urban poor because of lower prices. The evidence on this, however, is somewhat mixed. Minten's (2008: 785) work in Madagascar, for example, found that while global retail chains sell better-quality food, their prices are 40%–90% higher, *ceteris paribus*, than those seen in traditional retail markets. Battersby and Peyton's (2014: 153) research in low-income neighbourhoods in Cape Town found that supermarkets in low-income areas "typically stock less healthy foods than those in wealthier areas and, as a result, the supermarkets do not increase access to healthy foods and may, in fact, accelerate the nutrition transition". Riley and Legwegoh's (2014) comparison of food geographies in Blantyre, Malawi, and Gaborone, Botswana, sheds light on this debate. They note that the AFSUN surveys found a much higher rate of food security in Blantyre (48%) relative to Gaborone (18%), pointing out that this is surprising considering Blantyre's lower 'development' status in terms of urban infrastructure, economic opportunities and urban planning. Riley and Legwegoh (2014: 52) argue the major difference between the two cities is that:

> The majority of households in the Blantyre survey produced some of their own food and usually purchased food from informal markets while most

of the food in Gaborone is produced outside of the country and accessed through international supermarket chains.

The evidence also doesn't suggest that the informal food retailers are automatically replaced by supermarkets. Lane *et al.*'s (2012) Botswana case study shows that despite the penetration of South African retailers, informal food retailing persists. Abraham's (2010: 115) analysis of the implications of expanded supermarket presence in Zambia shows informal food markets are "progressively more resilient and competitive, despite the growth of supermarkets". Peyton *et al.* (2015: 36) trace the penetration of supermarkets in Cape Town, South Africa, and show that while supermarkets have been successful in penetrating some low-income communities, "they are often incompatible with the consumption strategies of the poorest households, revealing the significance of the informal economy in Cape Town."

The policy and regulatory environment for informal retailers

There is a rich evidence base on the policy and regulatory environment for public space trading (Bhowmik 2010; Roever and Skinner 2016). Informal food vendors, while seldom given exclusive attention, are an important subgroup within this group of traders. These analyses give critical insights into the challenges faced. The evidence base on the environment for informal retailers operating from their homes in residential areas has received minimal attention.

Existing evidence suggests that exclusionary practices are commonplace. For example, analysis of the research policy network of the WIEGO (Women in Informal Employment: Globalising and Organising) news database shows that, over a three-year period, there were over 50 cases of significant eviction of street traders reported in cities across the global South. Notable cases in Africa include Cairo, Harare, Johannesburg, Lagos and Luanda, with high levels of violence noted in Cairo, Harare and Luanda. Exclusionary practices can be placed on a continuum. At the more extreme end there are large-scale violent evictions, where street vendors are simply removed from public spaces. In less severe cases some or all vendors are relocated, but often to more marginal locations with low pedestrian footfall and/or inadequate facilities. Finally, there is also ongoing lower level harassment of vendors by predatory state officials, often facilitated by legislation. Evidence from Africa on these different types of exclusionary practices and the rationale behind them will now be reviewed.

Zimbabwe's specific approach to the informal sector has drawn a lot of attention. Operation Murambatsvina in May 2005 saw the largest scale and possibly the most violent removal of informal housing and street vendors on the continent in recent history. UN-Habitat estimated that some 700 000 people in cities across the country lost either their homes, their source of livelihood or both (Tibaijuka 2005: 7). Sites where informal sector workers gathered to market their wares, as well as formal markets, some of which had been in operation for decades, were targeted. It has been suggested that urban areas (and informality in these areas) have been targeted as these populations have tended to vote for the political opposition. Tibaijuka (2005) estimates that one in every five Zimbabweans were affected. Political affiliations, although critical, can thus only be part of the rationale behind these actions. Kamete (2009: 897) argues that planners and the planning system served as "the handmaiden of state repression" during the operation. More recent evidence suggests that the intolerance of the Zimbabwean state to the informal sector in general and street trade in particular has continued (Human Rights Watch 2015; Rogerson 2016).

Zambia has also had a longstanding ambivalence to informality. Tranberg Hansen and Vaa (2004: 66–67) note how in 1999,

> council workers, police and paramilitary in riot gear razed the temporary market structures of Lusaka's city centre, extending the demolition the following night and weeks all across the city, into townships and residential areas. . . . In June, similar operations took place on the Copperbelt and in the towns along the line-of-rail.

Tranberg Hansen and Vaa (2004: 68) identify a local authority leadership change as a key reason for these evictions. They note that a new mayor and council members had come to office in Lusaka and they were "bent on cleaning up the capital".

Riley's (2014) documentation of Operation Dongosolo, in which thousands of street traders were removed in Blantyre, Malawi, in 2006, also highlights political factors, pointing particularly to the formation of the Democratic People's Party (DPP) as the governing party and associated shifts in patronage networks. He also suggests that the notion that "the poor do not belong in the city" was a key factor (2014: 443). Riley's study considers the implications of removing traders for urban food security, and in doing so argues:

> Dongosolo reshaped the geographies of where people could buy food and where they could earn a living. It re-established the primacy of formal-sector businesses and middle-class lifestyles, which served both contingent political purposes and long-standing expectations of what urban space should look like.
>
> (Riley 2014: 443)

Brown *et al.* (2014) investigated street vendor removals in Dar es Salaam in 2006–2007 and Dakar in 2007. They analysed the political initiatives behind the clearances and the property-rights regimes underpinning them. They particularly focused on the different roles of social movements that resulted in emerging political power in one city and passive marginalisation in another.

Another strand of research examines the actors behind the design of evictions and relocations. Many of these analyses point to the role played by property developers and management companies in the exclusion of vendors. Morange's

(2015) analysis of the displacement of street vendors in Nairobi's Central Business District, for example, details the role of private-sector actors. In explaining the disappearance and/or relocation of the periodic bazaars that were once such a key feature of Istanbul, Öz and Eder (2012: 297) note that this has been driven by rising real estate prices in the city, where "land has simply become too precious a commodity to be left to the bazaars". Lindell and Appelblad (2009) trace the privatisation of the management of city markets in Kampala, Uganda, and the implications for vendors and their associations, arguing that private interests external to the markets are taking over the management functions and side-lining or even repressing vendors' associations.

On planning practices, Kamete and Lindell (2010: 889) argue that many urban governments in sub-Saharan Africa have abandoned comprehensive planning and increasingly resort to ad hoc 'sanitising' measures. In assessing removals in Harare and Maputo, they argue that underlying factors included the political strategies of elites seeking to maintain and strengthen political control over urban areas, rationalising and legitimising such unpopular interventions by appealing to ongoing efforts at 'city marketing' through international events and referring to the imperative of upholding 'a modern city image'.

More common than large-scale evictions, however, is a type of low-level harassment of street traders that stems in part from uncertain policy and legal environments. Roever and Skinner (2016), drawing on research conducted in five developing country cities (including Durban, Accra and Nakuru) to examine the more 'everyday' challenges that street vendors face, demonstrate the livelihood impacts of generalised workplace insecurity, harassment and confiscation of merchandise on street vendors' earnings, assets and time. This type of everyday harassment typically requires vendors to pay bribes to local authorities and subjects them to confiscation of merchandise. A group that is particularly vulnerable to this are foreign street traders and spaza shop owners (see Crush *et al.* 2015). Where licences are issued, the number of vendors considerably exceeds the number of licences. Lyons and Snoxell's (2005: 1078) study of Nairobi, Kenya, found that 7000 licenses and formal sites were offered to an estimated 500 000 street traders. Kamunyori (2007: 33) reports that, in Nairobi, council inspectors make several times their monthly salaries from bribing street traders. She records the monthly salary of these so-called *askaris* as approximately US$50. This points to a more systemic problem: until local officials in African cities are better paid, this kind of corruption will be difficult to root out.

In many countries, there is a hostile legislative environment. The news analysis shows that new laws banning street trading were reported in Angola, Jordan, Mali, Malawi, Nigeria and Zambia. A recent development is not only placing a ban on street vending, but also declaring purchasing from street vendors to be a criminal offence. Instances of this have been reported in Malawi, Nigeria and Zambia. The research that has been conducted on legislative frameworks that apply to informal retailers operating in residential spaces shows how these are inappropriate and often destructive (see Ahmed *et al.* 2015: 30–31, on Nairobi's informal settlements and Denoon-Stevens *et al.* 2017 on land use regulations and township businesses in South Africa).

In Africa there were only two cases that have been documented in which city councils integrated street traders into urban plans: Dar es Salaam, Tanzania (Nnkya 2006), and Warwick Junction in inner city Durban, South Africa (Dobson and Skinner 2009). Both are dated in that there have been subsequent reversals on these more progressive approaches (see Skinner 2010).

Conclusion

This chapter has shown the importance of the informal sector in food security. Despite this, the review of policy and practice trends for informal retailers paints a picture of widespread evictions, relocations and hostile regulatory and political environments. This has serious food security implications that are seldom recognised or assessed. The literature provides a rich framework of reference, both empirically and theoretically, on processes of exclusion. It suggests that, in understanding exclusion, attention should be paid to city aspirations, the role of the private sector (particularly property developers and management companies), the role of party political processes, and underlying planning approaches and legal frameworks. There is, however, a surprising dearth of work on whether alternative planning approaches could be applied to planning not only *for* but also *with* the informal economy and alternative laws.[8] Since policy and practice are frequently the upshot of every political struggles, ways to strengthen collective voice and action among informal operators is a critical area for action.

Notes

1 It was only in 2003 that the International Conference of Labour Statisticians (ICLS) outlined definitional norms. Vanek *et al.* (2014:6) summarises these norms as follows:

> The '*informal sector*' refers to unincorporated enterprises that may also be unregistered and/or small; '*informal employment*' refers to employment without social protection through work both inside and outside the informal sector; and the '*informal economy*' refers to all units, activities, and workers so defined and the output from them.

Prior to 2003, the terms 'informal sector' and 'informal economy' were used interchangeably, often suggesting a difference in focus, with those wanting to emphasis the links between the formal and informal opting to use the term 'informal economy'. This chapter and the data in it are informed by the ICLS definitions and is focused on the *informal sector*, so uses this term throughout.

2 Given country specificity as to what is considered urban, 'non-agricultural' work is used as a proxy for 'urban'.

3 These reflect in 'informal employment' rather than informal sector statistics.

4 These figures do not apply to the whole city but just to the low-income neighbourhoods surveyed.

5 Percentages represent the number of cases reporting using a food outlet type. Since any one household uses multiple food outlets, the percentages add up to more than 100.

6 Fieldwork was conducted in Epworth in December 2016, in Kismu in January and February 2017 and in Kitwe in March and April 2017. The samples in Epworth and Kismu

are representative for these cities. However, in Kitwe the survey was conducted in two selected neighbourhoods. Similarly, the AFSUN data was gathered in selected neighbourhoods. Due to different timing of the fieldwork and different sampling strategies, the data is not strictly comparable but should be regarded as indicative.

7 With grateful acknowledgement to Issahaka Fuseini for data analysis.

8 In March 2014, the Indian Parliament ratified the Protection of Livelihood and Regulation of Street Vending Act, providing legal protection for street traders and affirming that street hawking is a fundamental right when carried out in designated spaces. The new Act also requires recognition of 'natural markets' where street traders have congregated in response to dynamic local demand and requires that town vending committees, which have representation from vendors, be established. The content and implementation of these new approaches need to be analysed.

References

Abrahams, C. (2010). Transforming the region: Supermarkets and the local food economy. *African Affairs*, 109/434, pp. 115–134.

Ahmed, S., Simiyu, E., Githiri, G., Sverdlik, A. and Mbaka, S. (2015). Cooking up a storm, community-led mapping and advocacy with food vendors in Nairobi's informal settlements. *Working Paper*, IIED, London.

Battersby, J. and Peyton, S. (2014). The geography of supermarkets in Cape Town: Supermarket expansion and food access. *Urban Forum*, 25, pp. 153–164.

Bhowmik, S. (2010). *Street vendors in the global urban economy*. New Delhi: Routledge.

Brown, A., Msoka, C. and Dankoco, I. (2014). A refugee in my own country: Evictions or property rights in the urban informal economy? *Urban Studies*, 15(12), pp. 2234–2249.

Chicho-Matenge, N. and Nakisani, C. (2013). An assessment of challenges faced by micro-enterprises in Botswana: A case of street food vendors in Gaborone. *International Journal of Learning and Development*, 3(5), pp. 57–73.

Crush, J. and Frayne, B. (2011). Supermarket expansion and the informal food economy in Southern African cities: Implications for urban food security. *Journal of Southern African Studies*, 37(4), pp. 781–807.

Crush, J., Chikanda, J.A. and Skinner, C. (eds.) (2015). *Mean streets: Migration, xenophobia, and informality in South Africa*. Southern African Migration Project (SAMP), African Centre for Cities (ACC) and the International Development Research Centre (IDRC), Cape Town.

Denoon-Stevens, S., Charman, A., Tonkin, C. and Demeestére, R. (2017). A case of Kafkaesque complexity: The obstacles of land use management on township microenterprise formalisation. Cape Town: Sustainable Livelihood Foundation. Available at: www.livelihoods.org.za/wp-content/uploads/2017/03/A-Case-of-Kafkaesque-Complexity-Draft-Paper.pdf [Accessed 15 Nov. 2017]

Dobson, R. and Skinner, C. (2009). *Working in Warwick: Including street traders in urban plans*. Durban: University of KwaZulu-Natal, School of Development Studies.

Duflo, E. (2003). Grandmothers and granddaughters: Old-age pensions and intrahousehold allocation in South Africa. *World Bank Economic Review*, 17(1), pp. 1–25.

Human Rights Watch, (2015). Zimbabwe Violent Crackdown on Street Vendors, 29/07/2015. Available at: www.hrw.org/news/2015/07/29/zimbabwe-violent-crackdown-street-vendors [Accessed 28 Feb. 2018].

Humphrey, J. (2007). The supermarket revolution in developing countries: Tidal wave or tough competitive struggle? *Journal of Economic Geography*, 7, pp. 433–450.

International Labour Organisation. (2013). *Women and men in the informal economy: A statistical picture*, 2nd ed. Geneva: International Labour Office.

Kamete, A. (2009). In the service of Tyranny: Debating the role of planning in Zimbabwe's urban 'clean-up' operation. *Urban Studies*, 46(4), pp. 897–922.

Kamete, A. and Lindell, I. (2010). The politics of 'non-planning' interventions in African cities: Unravelling the international and local dimensions in Harare and Maputo. *Journal of Southern African Studies*, 36(4), pp. 889–912.

Kamunyori, W. (2007). A growing space for dialogue: The case of street vending in Nairobi's Central Business District. Master's thesis, Department of Urban Studies and Planning, Boston, MA: Massachusetts Institute of Technology.

Lane, K., Hovorka, A. and Legwegoh, A. (2012). Urban food dynamics in Botswana: Insights from Gaborone's Central Business District, *African Geographical Review*, 31(2), pp. 111–125.

Levin, C., Ruel, M., Morris, S., Maxwell, D. and Armar-Klemesu, M. (1999). Working women in an urban setting: Traders, vendors and food security in Accra. *World Development*, 27(11), pp. 1977–1991.

Lindell, I. and Appelblad, J. (2009). Disabling governance: Privatisation of city markets and implications for vendors' associations in Kampala, Uganda. *Habitat International*, 33, pp. 397–404.

Minten, B. (2008). The food retail revolution in poor countries: Is it coming or is it over? *Economic Development and Cultural Change*, 56(4), pp. 767–789.

Morange, M. (2015). Street trade, neoliberalisation and the control of space: Nairobi's central business district in the era of entrepreneurial urbanism. *Journal of Eastern African Studies*, 9(2), pp. 1–23.

Nnkya, T. (2006). An enabling framework? Governance and street trading in Dar es Salaam, Tanzania. In: A. Brown, ed., *Contested space: Street trading, public space and livelihoods in developing cities*. Warwickshire: Intermediate Technology Publications, 79–98.

Otoo, M., Ibro, G. Fulton, J. and Lowenberg-DeBoer, J. (2012). Micro-entrepreneurship in Niger: Factors affecting the success of women street food vendors. *Journal of African Business*, 13(1), pp. 16–28.

Öz, O. and Eder, M. (2012). Rendering Istanbul's periodic bazaars invisible: Reflections on urban transformation and contested space. *International Journal of Urban and Regional Research*, 36(2), pp. 297–314.

Peyton, S., Moseley, W. and Battersby, J. (2015). Implications of supermarket expansion on urban food security in Cape Town, South Africa. *African Geographical Review*, 34(1), pp. 36–54.

Reardon, T., Timmer, P. and Berdegue, J. (2004). The rapid rise of supermarkets in developing countries: Induced organizational, institutional, and technological change in agrifood systems. *Journal of Agricultural and Development Economics*, 1, pp. 68–83.

Riley, L. (2014). Operation Dongosolo and the geographies of urban poverty in Malawi. *Journal of Southern African Studies*, 40(3), pp. 443–458.

Riley, L. and Legwegoh, A. (2014). Comparative urban food geographies in Blantyre and Gaborone. *African Geographical Review*, 33(1), pp. 52–66.

Roever, S. and Skinner, C. (2016). Street vendors and cities. *Environment and Urbanisation*, 28(2), pp. 359–374.

Rogerson, C. (2016). Responding to informality in urban Africa: Street trading in Harare, Zimbabwe. *Urban Forum*, 27(2), pp. 229–251.

Skinner, C. (2010). Challenging city imaginaries: Street traders struggles in Warwick junction. *Agenda*, 81, pp. 101–109.

Skinner, C. and Haysom, G. (2016). The informal sector's role in food security: A missing link in policy debates? *Working Paper 44*. Cape Town: PLAAS, UWC and Centre of Excellence on Food Security.

Steyn, N., Mchiza, Z., Hill, J., Davids, Y., Venter, I., Hinrichsen, E., Opperman, M., Rumbelow, J. and Jacobs, P. (2013). Nutritional contribution of street foods to the diet of people in developing countries: A systematic review. *Public Health Nutrition*, 17(06), pp. 1363–1374.

Tibaijuka, A. (2005). Report of the fact-finding mission to Zimbabwe to assess the scope and impact of Operation Murambatsvina by the UN Special envoy on human settlements issues in Zimbabwe. Available at: www.un.org/News/dh/infocus/zimbabwe/zimbabwe_rpt.pdf [Accessed 5 May 2017].

Tinker, I. (1997). *Street foods: Urban food and employment in developing countries*. Oxford: Oxford University Press.

Traill, W.B. (2006). The rapid rise of supermarkets? *Development Policy Review*, 24(2), pp. 163–174.

Tranberg Hansen, K. and Vaa, M. (2004). *Reconsidering informality: Perspectives from urban Africa*. Uppsala: Nordic Africa Institute.

Vanek, J., Chen, M., Carré, F., Heintz, J. and Hussmanns, R. (2014). Statistics on the informal economy: Definitions, regional estimates and challenges. *WIEGO, Working Paper (Statistics) 2,* Manchester, United Kingdom.

Wills, G. (2009). South Africa's informal economy: A statistical profile. *WIEGO, Working Paper* 7, Manchester, United Kingdom.

8 Planning and governance of food systems in Kisumu City

Patrick Odhiambo Hayombe, Fredrick Omondi Owino, and Frankline Otiende Awuor

Introduction

This chapter provides an overview of the food system of Kisumu City and how it has been shaped by planning and governance. In this chapter, the food system is understood as being all the processes food undergoes from production to consumption. It entails growing the food, harvesting, processing, storage and distribution, retailing, consumption, and waste management. In addition, it also includes value addition, preservation, packaging, safety checks, marketing, pricing, and disposal (Bakker *et al.* 2008).

Given Consuming Urban Poverty's (CUP) concern with the linkages between the food system and food security, it is essential to locate this work within a food security framework. This is because if there is a crisis in any part of the food system, it manifests in urban food insecurity and poor health. The work presented in Chapter 16 of this book indicates high levels of food insecurity within Kisumu, which confirms Mireri *et al.*'s (2007) work that found food poverty levels as high as 53.4% in the city. Findings that the Kisumu food system does not ensure food security to over half its inhabitants suggest that it is a particularly weak food system.

This chapter considers the role food governance plays in shaping Kisumu City's food system. Food governance is here defined as "the range of political, organizational, and administrative processes through which stakeholders (including citizens and interest groups) articulate their interests, exercise their legal rights, take decisions, meet their obligations, and mediate their differences" (Bakker *et al.* 2008: 1894). The chapter focuses on how Kisumu City's food system is shaped by planning and governance at the national and local levels. In particular, it looks at how national and local laws, policies, plans, programmes, by-laws, and extra-governmental actors can shape the food system of a secondary city like Kisumu. This chapter is informed by content analysis of field observations, secondary documents, and in-depth interviews and descriptive statistics of data from household, retailer, and five food value chain surveys conducted in Kisumu City. These data were collected from February 2016 to February 2017.

Historic and current national policy influence on Kisumu City's food system

Kisumu City's current food system should be understood as having been formed in part by colonial and post-colonial governance interventions. The following subsections are a brief overview of these interventions.

Food policy

There has been little explicit focus on food in Kenya's cities by the government. This has its roots in colonial governance structures, which framed the food question as a commercial agricultural question, which therefore meant that it had no urban mandate. The Devonshire White Paper of 1923 allocated "high potential areas" for white settlements and the "low potential areas" for native settlement, thus privileging large-scale commercial agriculture for colonial export. The land allocated for native settlement was mainly for subsistence and pasture production. This was reinforced by the Swynnerton Plan of 1955 which sought to increase Kenya's cash-crop production through improved markets and infrastructure, the distribution of appropriate inputs, and the gradual consolidation and enclosure of land holdings (Anderson 2005). Even though food and cash-crop production increased in the country, urban areas still remained food insecure (Guyer 1987). Further complicating and thus exacerbating the issue was that fact that, under the Swynnerton Plan, Kisumu town fell under native land but was also designated as an urban centre for commercial and residential neighbourhoods.

At independence in 1963, the Sessional Paper No. 10 on African Socialism and Its Implication to Planning in Kenya (Republic of Kenya 1965) endeavoured to spur the development of Kenya to be self-sufficient. In doing this it aimed to address food security issues through increased agricultural production and economic access. It was stipulated that production would be improved through provision of quality plant and animal seeds, land consolidation, extension services, development credit, provision of agricultural supplies like fertilizers, and training in modern technologies. Agricultural produce boards were established to process and market farm produce, thereby providing income to farmers. The gained income would be used to economically acquire other food items not produced by the farmer in question. Food price controls were exercised to ensure that a majority of people in the country, especially the urban poor, had access to food (Gow and Parton 1992). Additionally, food exports were restricted and imports substituted to minimize spending that would be used to develop the country's food system. This shift in policy approach impacted the Kenyan urban food systems in various ways. For instance, it saw Kisumu obtain a grain storage facility provided by the National Cereals and Produce Board (NCPB).

In the 1980s, food and agricultural markets experienced liberalization. This led to the increased export orientation of the food system. Consequently, this meant that in Kisumu the main sources of protein, namely the fish Nile perch (*Latesniloticus*) and tilapia (*Oreochromisniloticus*), become less available for local markets. Thus, the city's residents were obliged to depend on rejects or juveniles or consume more silver fish or sardine (*Rastrineobola argentea*). Subsequently, traditional fish-smoking kilns have disappeared from Kisumu's food system, giving way to modern fish processing factories (Abila and Kisumu 2003). Some positive impacts of the increased fish exportation include the modernization of fish landing beaches in the city and improved hygiene in fish handling.

Urban policy

Just as food policies have not explicitly addressed, but have profoundly affected, urban food systems, urban planning and policy has equally been silent on food systems issues. In Kenya, "Urban Planning" as a governance tool can be traced back to the 1931 Town and Country Ordinance, the Town Planning Act of 1948, and the Land Planning Act of 1961. These Acts were geared at controlling the development of urban and rural land use, but they failed to adequately address urban food system issues. These three pieces of legislation were repealed as urban planning and governance tools in 1996 by the Physical Planning Act (PPA) Cap 286. Sections 16 and 24 of the PPA provide for the preparation of regional and local physical development plans respectively. Subsequently, certain urban areas have been designated as markets where stalls and shops can be constructed while other areas have been designated as industrial areas for wholesale outlets for food and non-food items. Although markets are explicitly referred to, there is no specific mention of food within the Act.

One way in which urban planning specifically engages food is through the by-laws that prohibit urban agriculture. These colonial-era by-laws are still in place today, even though the city's borders have since expanded to include rural areas. As such, almost 80% of the city's land is rural and predominately used for subsistence farming – a technical violation of the city's by-laws (Mireri *et al.* 2007; Mr. Aluoch Orombe, personal communication, 25 May 2016). Despite the fact that these laws are rarely enforced, even within more urban areas of the city (Mireri *et al.* 2007), this continued prohibition of urban agriculture suggests a framing in which the rural is imagined as the site of production and the urban solely as the site of consumption. This reflects a wider ambivalence towards the governance of food at the urban governance scale.

Kenya Vision 2030

Kenya's current strategic development plan, Kenya Vision 2030, has seen a number of flagship projects started in the country. Although not explicitly focused on the food system, these projects are reshaping many aspects of the food systems in various direct and indirect ways. For example, one project focuses on the improvement of road networks connecting Kisumu City with other towns and cities in the country. This has notably reduced the time and increased the physical ease with which food is transported into the city, as these roads were formerly in bad condition and too narrow for fast and safe transportation of food. Another example within Kisumu City is the expansion and improvement of storm water drainage infrastructure which has improved market hygiene and food safety.

Constitutional provisions

Article 43 in Kenya's new constitution (2010) provides for the right to food. This provision has seen county governments seek to ensure food security in their areas of jurisdiction in various ways. For example, Kisumu County is in the processes of developing an Agriculture and Food Security Bill. Additionally, the new constitution has led to the establishment of county governments through the County Governments Act of 2012. This Act, together with others such as the Land Act of 2012, the Urban Areas and Cities Act of 2011 (Amendment 2016), and the Physical Planning Bill (2015), has enabled Kisumu County and City to prepare spatial and other plans through their own relevant departments and boards. The Kisumu County Integrated Development Plan (ISUD) (2013–2017) is one such plan. While this plan has various food system implications, it also does not explicitly identify food as an object of planning.

How Kisumu City's food system is shaped by planning and governance

Production, harvesting, and storage

Kisumu City currently has no policy on agricultural production, harvesting, and storage. The existing Agricultural Act of 2012 governs land use in the city's rural areas. As noted above, existing by-laws in the city prohibit urban agriculture. Whereas there is limited crop and livestock keeping within the urban areas of the city, the practice is common in the rural unplanned areas of the city. The CUP survey data reveal that less than 15% of Kisumu City residents grow some of their own food and only 26% keep some livestock. The production of vegetables and grains that does take place in Kisumu City is largely for subsistence purposes and involves minimal storage or processing. Observations from the field indicate that commercial production of chicken and indigenous vegetables is increasing to meet the increased food demand of the city. However, this is not well supported by the city or county government in terms of infrastructure, market provision, or funding.

Kisumu City is a net food importer due to the effect of the long-term export orientation of the food and agricultural sector in Kenya. Moreover, it is impacted by the city's rapid population growth which has increased demand well beyond the productive capacity of its peri-urban areas. Food imports are not only sourced from other counties in Kenya (Figure 8.1), but also from other countries, like Uganda. Kale and cabbage are imported from Kericho and

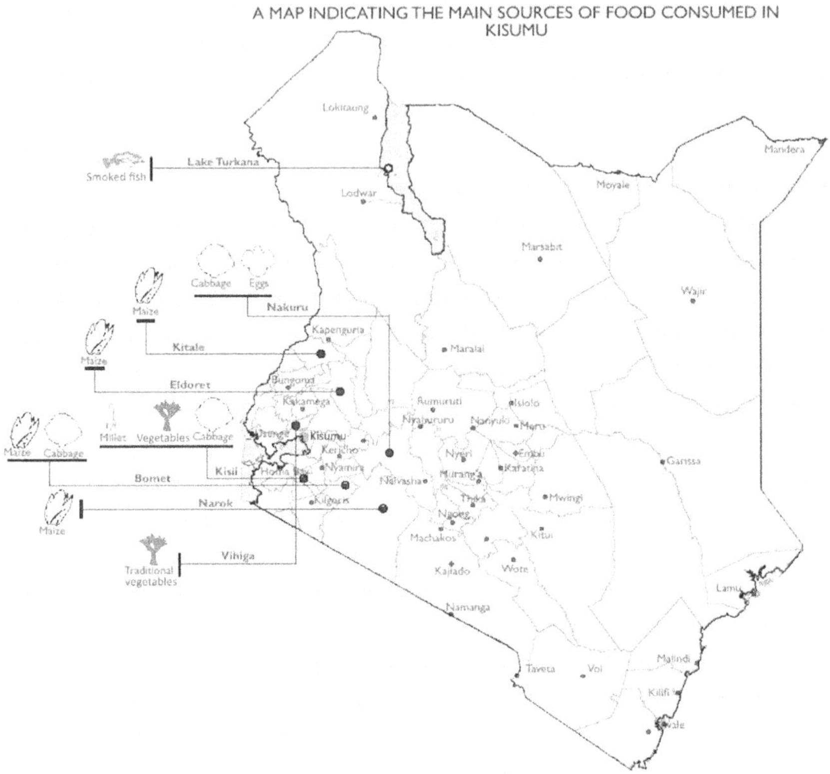

Figure 8.1 Kisumu City's sources of food as established via 2017 field survey (Authors' map)

Bomet, milk from Nandi and Kakamega, while some fish and eggs are imported from Uganda, making them more expensive in Kisumu. While fish is one of the staple proteins in the city, its supply from local fishing does not meet the local demand (Mairura 2010). The adoption of cage fish farming and other fish farming techniques have been promoted locally to meet this demand, although this has not had much traction as the cultural preference is to catch fish from the lake. It has also meant that fish from China have been brought into the market to supplement local supply. The sources of food consumed in Kisumu are discussed further in Chapter 13.

Food processing and preparation

County governance plays an important governance role within the food processing and preparation components of Kisumu City's food system.

Food processing in the city is largely small scale and farmer or retailer based, with the exception of fish for export, which is commercially processed in the city at Peche Foods fish factory. Fish and other animal foods for local consumption are usually processed (i.e. the removal of scales and innards from fish and chickens) by farmers and retailers. This is made possible in Kisumu's fish market by the availability of sinks for cleaning and processing. Most chickens are, however, processed at home instead of the market, and some are even further processed by deep frying, not only to enable them to last longer but also to prepare them for consumption. Sukuma wiki (*Brassica oleracea*) and some indigenous vegetable leaves are often destalked and sliced as a means of processing for sale. In these processing processes, public health officials, as empowered by the Public Health Act Cap 242 (Revised 2012), are pivotal in ensuring food safety for the public.

Another way in which Kisumu City governs its food system is through the regulation of food safety by the city's Department of Public Health. Those who process and prepare food for consumption in canteens, hotels, and restaurants are required by law to undergo medical check-ups for food-borne diseases like typhoid, amoeba, and tuberculosis. A number of tests are carried out but only urinalysis and stool microscopy are required for licensing. Another test that is technically required by the Department is for hepatitis B, but this is never done because of the cost implications. The operation of food outlets such as markets, slaughter houses, eateries, restaurants, cafes, fresh food kiosks, and stalls are required to conform to the Public Health Act (Revised 2012), Food, Drugs and Chemical Substance Act (Revised 2012), Meat Control Act Cap 356 (Revised 2012), Animal Diseases Act Cap 364 (Revised 2015), Fisheries Act Cap 378 (Revised 2012), and Dairy Industry Act Cap 336 (Revised 2012). To ensure compliance, food outlets are routinely checked for cleanliness and availability of sanitary facilities, clean water, and general health infrastructure. While these checks are supposed to cover the whole market, limited staff and equipment means that regulation tends to focus on food outlets as perceived hot spots for the spread of food-borne diseases.

Food safety surveillance is implemented in Kisumu according to certain provisions, for example an aflatoxin management programme and Anti-Microbial Resistance (AMR) in meat value chains. Inspection services are carried out by government regulatory agencies such as the Department of Veterinary Services, Department of Livestock Production, Department of Fisheries, Kenya Bureau of Standards, Kenya Plant Health Inspectorate Services (KEPHIS), National Public Health Laboratory Services (NPHLS), Government Chemist, and National Biosafety Authority (NBA), among others. Government regularly performs import control activities and domestic food control activities including certification (Oloo 2010).

Transportation

Kisumu's food system depends on a range of transportation means. Supermarkets use vans and trucks to transport stock from their suppliers to their branches in the city. Trucks deliver food to major markets in town. The small-scale informal sector depends either on foot transportation or an array of informal or formal public transport. Marginal businesses depend on food transportation with wares carried by means of plastic containers (basins or baskets), woven baskets, sacks, or milk containers. Less marginal businesses use bicycles, motorbikes, or tricycles – often in the form of boda-boda bicycle taxis – or travel with their wares on buses. While there are no specific food transport requirements within the Public Health Act – outside of the assertion that transport should be safe – the state plays an important role in shaping the transport system that in turn impacts upon the food system. As such the quality of road infrastructure and the regulation of public transport play an important role in Kisumu's food system. It is therefore important to consider the nature of transport infrastructure investments within Kenya Vision 2030 and development plans funded by external agencies.

Marketing and retailing

The 2013 ISUD for Kisumu (Nodalis 2013) identifies three metropolitan markets: the Fish Market and Kibuye Market, which operate as wholesale and retail markets, and Jubilee Market, which operates as a retail market. Sixteen other urban markets are identified, with six of these identified for partial or total demolition based on their limited economic importance. A further ten rural markets that serve the needs of the peri-urban population of Kisumu were listed. There also exists several informal neighbourhood markets (often adjacent to main arterial roads), as well as street traders and mobile vendors. Of these retailers, the CUP survey established that 28% of them trade on the street edge and 8% as mobile vendors, disregarding city by-laws.

Designated markets have been identified by the city government as the primary legal retail sites for vendors. Within these markets, the city collects daily rates from small-scale traders as a form of business permit to use the space and cleanliness infrastructure. However, there is not sufficient space to accommodate the small-scale traders. Consequently, many operate outside these formal markets. In the recent past, this led to the illegal occupation of a green park (Oile Park) in the city. This park has since been reclaimed but amid fierce protests from the occupying traders. The CUP findings indicate that traders traded outside designated markets in order to compete for customers. City health officials objected to this practice, indicating that when food items are displayed on pavements they can easily get contaminated with dirt from wind and passing wayfarers. Furthermore, health officials suggest that by so doing they obstruct pedestrians who often risk their safety as they are forced to walk on the roads.

An interview with the city's Finance Department, and corroborated by observations, established that taxes are collected from street traders in a bid to collect more revenue even though they trade on undesignated food retailing spaces. It was also observed that some of these are not receipted and are consequently given as bribes. As a result, many traders operate only in the evenings in an effort to evade authorities.

Planning of future market places should be informed by an understanding of the economic and spatial logics of small traders. There has been an increase in the construction of roadside stalls ('kiosks') as market spaces by private developers. They are, however, largely used for non-food business.

Waste management

The city government plays an important role in the governance of waste management, which is a key component of the food system. Consequently, the city attempts to ensure that waste is properly managed and disposed. However, proper waste management is hampered by the city's insufficient resources and a "callous" culture of waste management by the city's residents, a culture which is common across Kenya (Okaka 2014). This reflects limited awareness of the impact of waste on public health, public aesthetics, and drainage systems.

Organic waste, much of it food waste, comprises about 63% of the waste at the city's main dumpsite, Kachok (Ombara *et al.* 2015). While there are efforts to compost manure from organic waste in the city, the city's waste management system is not designed to produce manure as an output for farmers in the peri-urban and rural areas of the city and county. The city's urban food system is therefore largely linear with waste being the end point of the food system, rather than moving towards a circular system in which waste is reincorporated into the productive cycle.

How Kisumu's food system is shaped by extra-governmental forms of governance

As explained above, there are multiple intersections between formal governance and the food system in Kisumu, many of which are not explicitly food focused. However, governance extends well beyond formal government structures. This section considers the impact of extra-governmental forms of governance on Kisumu's food system.

Market organizations

In Kisumu City, like most urban areas in Kenya, legal and illegal forces shape the local food system. Legal entities include trader or market associations and farmer or producer cooperatives while illegal ones include cartels. An example of these cooperatives is Jubilee Community Based Organization for informal traders. These legal entities shape Kisumu's food system by organizing services that the city cannot satisfactorily provide like cleaning of market spaces and toilets, planning for the storage and security of their goods, maintaining order in the markets, setting price ranges for their goods, resolving conflict among traders, and forming a united front to dialogue with the city management on issues of interest to them. They additionally leverage through their numbers to get cheap reliable transport and storage facilities like freezers for their goods.

Whereas legal market entities are often associated with positive impacts on the food system, cartels are frequently linked with negative ones. For instance, cartels prevent new entrants into the market as they fear that they may spoil business by selling their goods at lower prices, thus spelling losses to other traders. Thus, new entrants may opt to sell their goods at another market in the city or in another town with lower volumes of buyers, thereby risking losses from some of their produce going bad. Alternatively, middle men and women in the food industry may collude with other big players or key institutions to their gain but at a high cost to the consumer (Mr. Charles Okwiri, Secretary Jubilee Market Traders Association, personal communication, 25 May 2016). For example, the high maize prices realized in Kenya in May 2017 as a result of maize shortage was a consequence of cartel-like behaviour between millers and traders (Kisero 2017). This kind of behaviour limits the economic accessibility of food to the urban and rural population.

Supermarkets

The number of supermarkets has been increasing in Kisumu. Formerly, these outlets were only situated in the Central Business District, but they are now increasingly located in or near residential areas. For instance, Tumaini Supermarket has two outlets near residential areas: one is located along the ring road delineating the boundary between Milimani residential area and Nyalenda slums and the other at Kondele junction near Manyata informal settlement. Nakumat Supermarket has an outlet at Kachok junction, which is near Polyview residential area and Nyalenda slums. Choppies Superstore, formerly Ukwala Supermarket, recently opened an outlet at Nyamasaria, an area which is at the extreme edge of the city's urban area. The city's peri-urban areas are also witnessing the opening of smaller supermarkets like Desmart.

These outlets are changing the food system and the way in which consumers engage with retailers. They provide opportunities for consumers to access all their required food from one shop, instead of purchasing from multiple shops and stores. Moreover, supermarkets' supply chains are different from those of small retailers. They source their food items from supplies with produce of consistent quality and supply, unlike small-scale retailers, who most of the time will go to the local markets or farmers and source what they perceive to be of good quality and at a cheap price (Neven and Reardon 2004). Furthermore, they are making available food items like imported fruits (e.g. oranges, apples, grapefruit, pears, kiwi fruit) and nuts (e.g. walnuts, cashew nuts, almonds), among others, that would ordinarily not be available in local municipal markets and from small-scale food vendors in informal neighbourhoods.

Large food stores are similarly impacting on the cooked food system. While in the past meals were only taken in eateries, restaurants, and food kiosks, supermarkets have also evolved to provide the same. It is not uncommon to see people buy and carry home cooked food or eat it within supermarket cafes. Consequently, supermarkets and new international entrants like Kentucky

Fried Chicken compete with established food outlets and street cooked food venders for wealthier consumers.

According to Neven and Reardon (2004), the establishment of supermarkets will ultimately displace stalls and small retail shops. However, the CUP household survey found that 55% of households bought less than a quarter of the food they purchased from supermarkets and new international entrants. Therefore, it seems that the direct displacement of stalls and small retail shops is not yet a reality. However, the increased presence of the kinds of food sold by supermarkets may be shifting diets through increased exposure to highly processed and sugary foods, which may change the food profile of the city.

Development agencies

International development agencies are playing an important governance role in shaping Kisumu's food system. For example, the Swedish International Development Agency (Sida) in partnership with the World Bank and Agence Française de Development (AFD) are supporting infrastructural developments (roads, drainage channels, footpaths, and water and sanitation infrastructure, as well as the installation of floodlights) in Kisumu. It is hoped that these developments will improve accessibility, and thus make it easier to transport local produce to markets in the city. Improved sanitation will decrease disease outbreaks in the city. The Kisumu Urban Project, a 40-million-Euro AFD project, has as a core project objective the reconstruction of Jubilee, Kibuye, and Otonglo markets and the removal of hawkers from Kisumu's streets to the newly constructed markets. Through its Accelerated Value Chain Development Programme, USAID seeks to boost food security in six western Kenya counties including Kisumu. They hope to achieve this by assisting farmers to invest in dairy farming, growing of sweet potatoes, pineapples, groundnuts, and drought-resistant crops such as sorghum. This programme is expected to shape Kisumu's food system by increasing production in the city's suburban and rural areas. Development agencies therefore play an integral role in shaping the food system and the conditions shaping the accessibility and utilization of food by Kisumu residents.

Private land developers

Since the start of the county governance system in Kenya in 2013, county government headquarters have witnessed rapid urban growth. Kisumu City, as Kisumu County's headquarters, is no exception for the city has witnessed an exponential growth in commercial and residential developments (Abdallah 2016; Owino *et al.* 2017). This, in turn, leaves less land for food production. This is due to the fact that commercial and residential developments are more profitable than food production on small pieces of land. Local food production as part of the food system is therefore likely to become even less significant as urban development pressures increase. Owino *et al.* (2017) report that land for food production is being converted to commercial and residential use.

Conclusion

Kisumu's food system is shaped by spatial plans and governance at the national and local levels, which often do not have an explicit focus on urban food systems. Furthermore, market organizations and supermarkets play important roles in the local food system that support, influence, or impede the efficient function of a food system. In planning for the food system of a city, it is important therefore to include all these actors, as it has clearly been shown that the forces that influence a food system go far beyond the governance of a municipality, city, county, or national government. In Kisumu, local planning processes at the county and city levels give little direct attention to the city's urban food system and therefore its food security, and yet they must be understood to be powerful actors in shaping this system.

References

Abdallah, R. (2016). *Influence of automation of building plans approval process on performance of employees at the planning department of Kisumu city-Kenya*. Unpublished Masters Thesis in Project Planning Management, University of Nairobi.

Abila, R.O. and Kisumu, K. (2003). *Impacts of international fish trade: A case study of Lake Victoria Fisheries*. Rome: FAO.

Ombara, D, Nyambuga, C, Oloko, M (2015), *Urban Sustainable Development Goals (USDGs) Targets and Indicators: The Case of Kisumu*. [online] MISTRA HQ. Available at: https://www.mistraurbanfutures.org/sites/mistraurbanfutures.org/files/usdg-report-kisumu_0.pdf.

Anderson, D. (2005). *Histories of the hanged: The dirty war in Kenya and the end of empire*. London: Weidenfeld and Nicolson.

Bakker, A.B., Schaufeli, W.B., Leiter, M.P. and Taris, T.W. (2008). Work engagement: An emerging concept in occupational health psychology. *Work & Stress*, 22(3), pp. 187–200.

Gow, J. and Parton, K.A. (1992). The evolution of Kenyan agricultural policy. In: *36th Annual Conference of Australian Agricultural Economics Society*. Canberra: ACT (Australia), pp. 101–105.

Guyer, J.I. (1987). Introduction. In: J.I. Guyer, ed., *Feeding African cities: Studies in regional social history*. Bloomington, IN: Indiana University Press, pp. 1–54.

Kisero, J. (2017). *Maize price crisis shows Kenya is a Bandit Economy*. Available at: www.worldgrain.com/news/news%20home/LexisNexisArticle.aspx?articleid=2757513561&e [Accessed 15 May 2017].

Mairura, O. (2010). Kenya: Urban settlements and development profile. In: C. Gossop, ed., *ISOCARP, sustainable city/developingworld*. Netherlands: ISOCARP, pp. 20–34.

Mireri, C., Atekyereza, P., Kyessi, A. and Mushi, N. (2007). Environmental risks of urban agriculture in the Lake Victoria drainage basin: A case of Kisumu Municipality, Kenya. *Habitat International*, 31(3), pp. 375–386.

Neven, D. and Reardon, T. (2004). The rise of Kenyan supermarkets and the evolution of their horticulture product procurement systems. *Development Policy Review*, 22(6), pp. 669–699.

Nodalis, C. (2013). *Kisumu ISUD-plan—Part 1*. Kisumu: Agence Française de Development.

Okaka, F.O. (2014). Public perception and awareness of health risks of climate change: A potential for adaptation and mitigation in Kenya. *African Journal of Education, Science and Technology*, 1(4), pp. 150.

Oloo, J. (2010). Food safety and quality management in Kenya. An overview of roles played by various stakeholders. *African Journal of Food, Agriculture, Nutrition and Development*, 1(11), pp. 4379–4397.

Owino, F.O., Hayombe, P.O. and Agong, S.G. (2017). The effect of diminishing urban green spaces on environmental quality in Kisumu City. *International Journal of Geography and Regional Planning Research*, 12(1), pp. 10–23.

Republic of Kenya, (1965). *African socialism and its application to planning in Kenya*. Nairobi: Government Printer.

Republic of Kenya, (1996). *The physical planning act Cap 286 of 1996*. Nairobi: Government Printer.

Republic of Kenya, (2010). *The constitution of Kenya*. Nairobi: Government Printer.

Republic of Kenya, (2012). *Land act*. Nairobi: Government Printer.

Republic of Kenya, (2012). *Meat control act Cap 356*. Revised 2012 (1977). Nairobi: National Council for Law Reporting.

Republic of Kenya, (2012). *Dairy industry act Cap 336*. Revised 2012 (1984). Nairobi: National Council for Law Reporting.

Republic of Kenya, (2012). *Public health act Cap 242*. Revised 2012 (1986). Nairobi: National Council for Law Reporting.

Republic of Kenya, (2012). *Food, drugs and chemical substance act Cap 254*. Revised 2012 (1992). Nairobi: National Council for Law Reporting.

Republic of Kenya, (2015). *Animal diseases act Cap 364*. Revised 2015 (1965). Nairobi: National Council for Law Reporting.

Republic of Kenya, (2015). *Planning bill*. Nairobi: Government Printer.

Republic of Kenya, (2016). *Fisheries act Cap 378*. Revised 2016 (1991). Nairobi: National Council for Law Reporting.

Republic of Kenya, (2016). *Urban areas and cities act* (Amendment 2016). Nairobi: Government Printer.

Swynnerton, R.J.M. (1955). *The Swynnerton report: A plan to intensify the development of African Agriculture in Kenya*. Nairobi: Government Printer.

9 Planning and governance of food systems in Kitwe, Zambia

A case study of food retail space

Jane Battersby and Francis Muwowo

Introduction

Historically, local governments in Zambia have played a critical role in the food system with efforts to ensure food security for urban residents through their management of markets. However, this chapter argues that the centrality of the governance of food as integral to the management of markets has been lost. With market management now largely being viewed as regulation of activities, provision of infrastructure, and control of land, decisions that profoundly affect the food system are increasingly made without consideration of the food security impacts of these decisions. While food systems include production, processing, packaging, distribution, retailing, consumption, and waste (HLPE 2014), the focus of this chapter is the retail sector, the main point of interaction between consumers and the food system, which is a key focus of the Consuming Urban Poverty (CUP) project. As Chapters 14 and 17 illustrate, the informal food retail sector is essential to the food security of Kitwe's urban poor, but its ability to meet these food security needs is dependent both on the location of these businesses and on the quality of infrastructure they have access too.

This chapter argues that the importance of these retailers as a source of food security has not been recognized by the state. It therefore focuses on three distinct, but connected, types of food retail space in Kitwe and examines how they are impacted on by the policies and actions of local government and other governance actors. The chapter argues that current governance of food retail spaces is determined by a normative view of an idealized 'modern' city, the limited capacity or resources to achieve this ideal, and the need to curate a complicated set of relationships with traders, market associations, large-scale private capital, and the development sector. The chapter further argues that in their efforts to achieve these urban goals, the food delivery purpose of food retail space management has been lost. The chapter concludes that reinserting the food system outcome of governance and planning decisions could provide new ways for local government to plan their market management strategies.

The state's role in urban food system governance and food security

Historically, the state played a central role in Zambian urban food systems. The main legal basis for this governance was the 1937 Market Ordinance, also called the Lusaka Markets Act. This Act placed markets under the control and management of local authorities, regulating market buildings, what goods could be sold, when and at what maximum prices, as well as giving local authorities the right to inspect and grade goods. The Act further stated that "no person shall, in any public place within a radius of two miles from the centre of a market, sell any goods except in a market" (Government of Zambia 1997, Section 5, Para 5), thereby prohibiting street vending. As Scott (1985: 87) notes:

> These laws had legal and social implications: they had the effect of creating large buffer zones around all authorized markets where commercial transactions were illegal, thereby protecting sanctioned traders from unsanctioned domestic and foreign traders; and they gave government and the party authority to set prices on selected foods and convenience items, thereby weakening the profit objective among marketeers and promoting the idea of equal access to essential foods and convenience items.

The 1937 Act was a powerful tool of spatial and economic control, in which control of the food system in order to meet urban food needs was seen as central. However, it was designed and promulgated in a predominantly rural country, with small urban centres characterized by almost full employment within the formal sector. This historical system of governance of food in urban Zambia would be poorly aligned to current urban realities, including an increasing population, high unemployment, massive informal sector employment, and an increasingly dominant corporate sector shaping the food system. This chapter discusses how governance of urban food retail spaces in Kitwe functions within the context of changing urban demand. The impacts of these structural transitions on the food system and food poverty are discussed in Chapters 14 and 17.

The Markets Act was repealed in 2007 and replaced by the Markets and Bus Stations Bill, the draft form of which included clauses about price ceilings. Following an internal government review, these price control clauses were removed on the basis that they were "incompatible with the liberalised economy" (Committee on Local Governance, Housing and Chief's Office 2006: 6). In this statement, the primacy of market forces in delivering food security is made apparent as a principle of governance, and the historical role of urban markets as part of a food system designed to ensure urban food security was abandoned.

The urban is largely absent from Zambian food policy. Current national food policy no longer envisages a central role for local government in ensuring urban food security. National food policy has focused on the production of maize, a Farmer Input Support Programme, and the purchase of maize above market prices to sell to millers at below market prices in order to ensure national food security (Africa Lead 2014). This has not, however, guaranteed low prices to the urban poor, and so President Edgar C. Lungu announced in December 2016 that the government would liberalize the price of maize in order to break the power of the milling cartels (Mulenga 2016).

Food is equally absent from urban policy. The Urban and Regional Planning Act was passed in 2015 and is meant to establish principles for urban and regional planning (Government of Zambia 2015). It is notable that in this document food is only mentioned once (in the context of production). This means that local government sees no clear food system or food security role for itself, despite many core functions of urban government, such as control of markets, public health, transport, and solid waste management systems and infrastructure, influencing the characteristics of the food system. Kitwe City Council (KCC) consists of 28 elected councillors and works together with the District Administration to execute development programmes. It views itself as being responsible for the development of local policies, for "refuse collection, street cleaning and lighting, promotion of public health, maintenance and other services" (KCC, undated). Within the KCC's own framing of its responsibilities, following the national lead, there is no acknowledgment of their role in shaping the food system or food security (KCC, undated).

In the absence of clear urban food policy, urban food system governance is the unacknowledged outcome of the contestations and collaborations of local government with other actors in the urban system, including traders, market associations, and private capital and development agencies.

Urban food system governance

The core argument of this chapter is that the urban food system is being shaped by governance and planning interventions undertaken by local government which are not viewed as food system interventions. This is demonstrated through the local government's governance of the city's main market, Chisokone Market, which lies in the heart of Kitwe's Central Business District (CBD). It was constructed in the 1930s in accordance with the guidelines of the 1937 Markets Act. It was then refurbished in the 1970s, but by the mid-2000s it was acknowledged that the market was over-crowded and unsafe (Chabalengula 2009).

In light of the growth of Kitwe, and the growth of trading as a livelihood following the retrenchments in the mining sector (Hampwaye 2008), the market has grown well beyond its original site. It currently occupies 15 acres of land and is estimated to accommodate up to 10 000 traders. However, the officially gazetted area of the market is comprised of just 1.9 acres. The remainder is comprised of the ungazetted Chisokone B, C, D, Chisokone Kingston, and Chisokone furniture market sites on land that is zoned for formal commercial activity, sites which have been occupied since at least the early 1990s (Brinkmann 2014). Marketeers on ungazetted land have in practice gained legitimacy through paying daily market fees to the City Council.

The Council faces the critical challenge of how to manage the market and make the site safe for trade. It has taken two approaches in trying to achieve this: upgrading the market and relocating marketeers.

Market upgrading

In 2009 the US$1m Chisokone Market Improvement Project was launched, managed by the KCC, UN-Habitat, the Ministry of Local Government and Housing, and the Danish International Development Agency (DANIDA) (UN-Habitat 2009: 37). The Kitwe 2012–2016 Strategic Plan identified the re-planning of Chisokone, focusing on upgrading the market, as one of 14 key potential areas for investment (KCC 2012). The government has subsequently made various commitments to upgrading the market site, including both gazetted and ungazetted areas, and the current mayor has repeatedly called for improvements to be made, particularly with regard to sanitation (Ntambi 2017b). However, it seems that the local government response is more reactive than proactive, with efforts to upgrade and rehabilitate the market largely taking place after disaster events, such as a fire in 2012 and a roof collapse in 2016 (LusakaTimes 2016), which is when funding from central government has tended to be made available. However, conditions at the market have remained unsafe, with high risk of fire, poor sanitation, and roofing collapses.

Market relocation

In an effort to address the problems associated with Chisokone, the KCC has developed three modern urban markets outside the city centre, in Nakadoli, Buchi/Kamitondo, and Ndeke, using European Union (EU) funding (Songolo 2017: 29). The rationale for developing these modern markets was to decongest Chisokone, with plans to move wholesalers from Chisokone to Nakadoli in 2012 and again in 2017 (UKZambians 2012; Kalero 2017).

However, efforts to move traders have been strongly resisted by the marketeers. Although established in 2009, Nakadoli Market still has low occupancy rates. The new markets have been contested sites since their development, with accusations that the KCC has failed to adequately maintain them and that levies are collected by Council officials for their own benefit (LusakaTimes 2013). There have been counter accusations that traders have sub-let their stalls illegally and have failed to pay their market levies (Chazingwa 2014). In July 2017 it was reported that Nakodoli traders owed K1.5m (around US$150 000) to the Council in unpaid market levies (Kalero 2017).

Moving of street traders

Within the context of extreme overcrowding and high environmental risk at Chisokone Market, in January 2017 the KCC embarked on a programme of evicting street traders. It was announced that 600 vendors moved from street-side retailing positions in January 2017 would be accommodated within the market and a new K1.6m shelter was to be erected (Ntambi 2017a). Nevertheless, the CUP census of food retail in Kitwe, conducted in June 2016, found significant trade in food taking place outside the formal market spaces. Food trade was particularly concentrated outside the Shoprite supermarket, one street away from Chisokone Market. The removal of street traders was justified as being necessary to enable the Council to provide better infrastructure for them in support of their trading activities and to enable the KCC to increase revenue collection, which would improve its capacity to improve infrastructure (Ntambi 2017a). However, as of October 2017, the upgraded space at Chisokone had not been completed. Street trading in Zambia, as elsewhere in Africa, operates under conditions of severe precarity and is deeply political, with phases of repression often associated with changes in political leadership at the national and local government scales.

Interrogating governance decisions

The following section discusses the politics and economics of the governance decisions to improve the conditions in markets, relocate them, and remove street traders and marketeers.

Market upgrading

Despite the technical illegality of much of Chisokone Market, local government persists in its efforts to upgrade the market site. This is in part because of the fundamental belief that designated market spaces are the correct places for retail and therefore require support. To this end, the Council has an office managed by a market master and cashiers, who collect levies. There are also dedicated Council police who are meant to maintain order and provide security. The whole market is therefore tacitly legitimized by the state (Chabalengula 2009: 32).

This legitimization provides political agency to the marketeers. Historically, market associations (non-profit associations representing marketeers' interests) played an important role in advocacy, lobbying, and collective bargaining to improve market conditions (Larsen 2009: 34). These associations were and are aligned to different political parties, with the Zambia National Marketeers Association (ZANAMA) being aligned to the Movement for Multiparty Democracy (MMD) and Zambian Traders and Marketeers Association (ZATMA) to the Patriotic Front (PF) (Jongh 2013). However, when the Markets Act was replaced by the Markets and Bus Stations Act in 2007, it stated that third-party associations would not be allowed to operate from within markets (Larsen 2009: 32). This removed a degree of power from the Market Associations. However, these associations still operate and marketeers remain a relatively powerful constituency, aided by their membership of Alliance for Zambian Informal Economy Associations (AZIEA), which connects Kitwe marketeers to international advocacy networks and offers training in engaging with local government (Jongh 2013). In Kitwe other market associations are also active, including National Traders and Marketeers Association of Zambia (NATMAZ) and Association of Vendors and Marketeers (AVEMA) (Kalero 2018). Given the multiple and competing market associations there is rarely a single marketer voice, which reduces their potential impact.

The market levies and taxes that marketeers pay to local government are the basis for a kind of social compact. They feel that their payment of levies and taxes to the KCC's representative, the market master, legitimizes their use of the land and leads them to expect services and infrastructure to be provided (Chabalengula 2009: 36). However, marketeers withhold payment of levies when services are not delivered, thus hampering the ability of the KCC to provide adequate services (UN-Habitat 2009). The ongoing precarity of tenure of the majority of Chisokone marketeers and the relationships between local government and marketeers has hindered local government's ability to adequately improve market infrastructure, therefore making market relocation an attractive option.

Market relocation

Efforts to relocate marketeers from Chisokone have been largely unsuccessful. The failure of Nakadoli Market reveals various contestations and alliances in urban decision making in Kitwe. The failure is often portrayed as a disagreement between the Council and the marketeers regarding appropriate space for trade. However, this section argues that there is a deeper set of contestations regarding the vision for a future Kitwe and which groupings feel is able to facilitate the city's development trajectory.

The tension between the Council and the marketeers is fairly clear. The Director of the Department of Planning in Kitwe has written that there is a "[t]rend of newly constructed market infrastructure being turned into white elephants due to marketeers' resistance to paying market levies, resulting in continued informal vending and untapped economic benefits" (Songolo 2017: 31). According to this interpretation, the problem lies with the marketeers. However, the marketeer perspective is somewhat different. Many have refused to move from Chisokone to the new markets on the basis of a lack of customers at the new sites (Kachemba 2017) and have challenged the Council on its unilateral decision making on trading in the city. The President of AVEMA, Able Chikwa, has stated:

> Kitwe is not the home for one individual who should just wake up in the morning and make a decision for people to see that he is working. No, that is not the way things work. Today, you move street vendors from the streets, the following day you want to move tomato wholesale traders from Chisokone to Nakadoli market. Just imagine, before the wounds of removing people from the streets can be healed, another thing comes up. It is not

> the fault of marketeers that the Nakadoli Market is a white elephant, hence no one should force marketeers to go to Nakadoli Market.
>
> (Kalero 2017)

Chikwa's concerns are reflected in the findings of the retail census conducted as part of the CUP project. This found that the market space at Chisokone is profoundly connected to other urban functions that make it a viable retail space, such as wholesalers located around the periphery of the market and proximity to major transportation hubs. The newly designed markets lack this hard and soft infrastructure, suggesting that those designing the spaces have not paid sufficient attention to the functioning of markets as part of the wider urban system. The CUP food retail and food poverty surveys also demonstrated the importance of retail location in business success and the central role of Chisokone as both a wholesale source of food for neighbourhood markets and a retail source of food for Kitwe residents. The efforts to move the market have therefore failed to consider how the food system operates.

This then raises the question of why these markets have been constructed in the absence of clear economic and spatial rationales. This chapter proposes that three elements have combined to reinforce the ideal of a modern market: limited capacity, external finance, and a vision of a modernized city centre.

The 2009 UN-Habitat profile of Kitwe city found that the Department of Development Planning was under-capacitated, having only nine staff members and just "a few old paper maps, which are outdated and inadequate for effective development planning and monitoring in the city" (UN-Habitat 2009: 9). In a context of limited planning capacity and resources cities like Kitwe have become increasingly dependent on donor funding to meet their development objectives (UN-Habitat 2009). The EU has a long history of investment in building new markets, dating back to the 1990s. The 19.6 million Euro Urban Markets Development Programme, which started in 2003, focused on both providing physical infrastructure and capacity building in 11 markets in Zambia, including the three new markets in Kitwe. The programme aimed at addressing:

- The proliferation of street vending and informal trading,
- The inadequacy of the physical structures of markets, which was attributed to the absence of clear regulations related to physical and functional organisation, which, it was claimed, resulted in the markets being shunned by traders,
- The degradation of market environments due to inadequate refuse collection and disposal facilities, which affected food hygiene and the health of the general public,
- Deterioration of the quality of living due to poor location of markets within the urban fabric,
- The overall inefficiency of the distribution system, which did not offer any guarantees to consumers on the quality of goods and prevented the development of entrepreneur capacities relating to markets despite favouring strong price competition,

- The inadequate legislative system which, apart from being obsolete, given social trends in urban areas, contained contradictions between the laws and regulations, and
- A disorganised management system, which resulted in a failure to administer markets well.

(GMZ/EC 2003 in Kalemba 2011: 41–42)

The priorities of the programme clearly indicate that the physical and economic model developed under the Markets Act was no longer viable for the urban realities of Zambian cities, but that the state did not have the capacity to address the challenges. This therefore was thought to necessitate donor support in order to develop more appropriate market spaces. However, this model committed the Council to developing markets that were not viable given local conditions. While there is popular pressure from marketeers to upgrade Chisokone Market, funding for this is limited. The EU funding for new markets, with the possibility of introducing new management systems, has therefore been favoured instead.

The decision to replace Chisokone Market was further informed by the KCC's vision of a modern CBD characterized by formal economic activities. The 2012–2016 Strategic Development Plan calls for "re-planning in a bid to modernize and decongest the Central Business District and reduce pressure on infrastructure" (KCC 2012: 16). The ungazetted portion of Chisokone is on land zoned for commercial investment. One local government official has gone on record as stating:

> [Chisokone Market] is the only parcel of land remaining in the commercial business district. It is planned that putting up quality investment on that land such as a shopping mall would not only provide the council with more revenues but also beautify the city.
>
> (Chabalengula 2009: 40–41)

Although the Town and Country Planning Act would enable the city to rezone the non-gazetted areas from commercial plots to market usage, which would provide marketeers with more rights and improved trading sites, this is not being done because the land is the only prime commercial land remaining in the CBD (Chabalengula 2009: 43). The long-term vision therefore remains the relocation of most of the market and its replacement by formal commercial activities.

Moving of street traders

In the context of the extreme overcrowding of Chisokone and the failure to relocate marketeers to decongest the market, why did the KCC take a decision to move street traders from their places of business to the market? The KCC has a long running antipathy towards informal traders, common across the African continent (Kamete 2013). Street trading is technically illegal in Zambia and is framed in law as a public nuisance. It is argued, variously, to steal customers from formal businesses, to be undesirable due to vendors' non-payment of taxes, to cause traffic congestion, to be unsafe and cause disease, and to harbour criminals (Ng'anjo 1994). These views of street trading are reflected in public discussions of the most recent street clearances (LusakaTimes 2017a). The 2017 street clearances in Kitwe are far from unique, with various other street clearances being documented in Zambia, including Kitwe (e.g. Hansen 2004).

The treatment of street trade in Zambia has gone through various phases of repression and permission (Hansen 2010). However, since 2011 there has been a softening of government approaches to the sector, and local government has been compelled to ignore street vending unless adverse issues arise. Hansen (2010: 20) argues this inconsistency in treatment is due to differences in the view of street vending held by different segments of the state. However, it is essential to view the state's treatment of street traders not just as differences in the positions held by different segments of the state, but also as being informed by political and economic interests, and a need to be seen to be responsive to the needs of other urban stakeholders, including market associations, large private-sector investors, and donor agencies. There is a street traders' association, the United Street Vendor Foundation (USVF), but unlike the marketeer associations, this is not aligned to AZIEA and is politically weak (Jongh 2013). Marketeers remain an important political force because of the collective contribution of their market levies to the KCC's revenue. As elsewhere, marketeers have argued that street traders have an unfair competitive advantage because many are not paying market levies and are therefore able to charge lower prices (Hansen 2004: 66; LusakaTimes 2017b).

While the concerns of the marketeers at Chisokone may have played a role in the decision to move street traders, this chapter argues that the efforts to move them are informed more by the KCC's interest in projecting Kitwe as a modern, orderly city which is responsive to large-scale private-sector interests. Under this model, trading needs to be controlled and contained. Hansen (2004: 70), working in Lusaka, argues that earlier clearances of street vendors were closely linked to economic liberalization efforts, with the presence of street vendors making the cities less attractive investment propositions for potential international investors. In both the 2012 and 2014 street vendor clearances in Kitwe, Shoprite (a major South African supermarket company which has expanded rapidly into the Zambia food system) was identified as being negatively affected by street trade (LusakaTimes 2012, 2014). This was argued by the Minister of Sport, Youth, and Child Development to be particularly problematic as "it [Shoprite] contributes a lot of money to Government in taxes" (LusakaTimes 2012).

A number of shopping malls, featuring international retailers, have been built or are under construction in Kitwe. These are viewed as symbols of urban development and cosmopolitanism. At the opening of the Mukuba Mall, for example, President Lungu said that it serves the needs of a "growing segment of our urban citizenry that has evolved and become more cosmopolitan and global in their exposure and preferences and have even more disposable income and options" (Lungu 2015). In efforts to attract more of this kind of investment, it is considered important that Kitwe's city centre is viewed as modern and clean, i.e. free from street traders.

Within the framing of the various laws on street trade and the treatment of street traders by the KCC there is no consideration of the role of street trade in the food system, or of the service it offers in terms of food security for the urban poor. However, CUP and other research suggests a strongly symbiotic relationship between formal and informal retailers (Abrahams 2009), rather than competition, and a clear logic to the physical location choices of traders. As with the plans to improve or relocate Chisokone Market, a set of planning and management decisions have been made that impact the food system, without the food system implications of these decisions being considered.

Conclusion

The 1937 Markets Act was undeniably politically and economically exclusionary, but it was also a powerful tool of spatial and economic control, in which regulation of the food system in order to meet urban food needs was a central component. However, the conditions of the Act were unable to respond to the political, demographic, and economic shifts in Zambia's cities post-independence.

This chapter does not call for a return to the rigid state control of the food system attempted under the 1937 Markets Act. It has highlighted the ways in which the governance of food retail has shifted in response to changing urban economies and politics in Kitwe. Markets and street trade have been, and remain, intensely political. The chapter has identified significant local government powers to shape urban food retail, yet it has limited capacity to act – in terms of both human and financial resources. The over-arching policy and planning position of KCC favours modernization and a desire to remove or control the informal sector. In the absence of sufficient local government capacity, decisions about the location of food retail are informed by a set of actors, including market associations, donor agencies, and large-scale private-sector interests, all of whom provide financial resources and therefore increase the capacity of the KCC directly or indirectly. None of these stakeholders' actions are informed by the needs of the overall food system or the food security implications of their interventions.

The cumulative impact of an under-capacitated local government with no clear food mandate and governance decisions influenced by external actors is a highly contradictory set of governance responses that swing from tacit support for market traders to repression depending on the politics of the moment. Given the importance of the informal food retail sector to the food security of Kitwe's urban poor, governance practices that undermine the viability of these businesses undermine food security. It therefore appears that the governance priorities of the KCC and the other stakeholders who shape governance decisions are not well aligned with the interests of the city's urban poor. This suggests a need for more explicit policy attention to be paid to shaping the urban food system in the interests of achieving food security, implying changes to the current policies and practices of local government.

References

Abrahams, C. (2009). Transforming the region: Supermarkets and the local food economy. *African Affairs*, 109(434), pp. 115–134.

Africa Lead, (2014). *Zambia food security policy assessment: Institutional architecture for food security and policy change*. Available at: http://pdf.usaid.gov/pdf_docs/PA00K3GB.pdf [Accessed 23 Feb. 2018].

Brinkmann, F. (2014). *Kleinhandel unter schwierigen institutionellen Voraussetzungen—Reaktionen, Strategien und Lösungen der Händler auf dem Chisokone-Markt in Kitwe/Sambia*. PhD, Bayreuth International Graduate School of African Studies, University of Bayreuth.

Chabalengula, S.K. (2009). Towards incorporating the urban informal economic sector in city development by local authorities: The case of Chisokone market in Kitwe, Zambia. M.A, UMD Rotterdam.

Chazingwa, M. (2014). Nakadoli market stalls repossessed. *Times of Zambia*. Available at: www.times.co.zm/?p=1994 [Accessed 23 Feb. 2018].

Committee on Local Governance, Housing and Chief's Office, (2006). *Report on the markets and bus stations bill (NAB No. 8 of 2007) Referred to the Committee on Local Governance, Housing and Chief's Affairs for the First Session of the Tenth National Assembly Appointed on 8th November 2006*. Available at: www.parliament.gov.zm/sites/default/files/documents/committee_reports/Final%20Report%20local.pdf [Accessed 23 Feb. 2018].

GRZ/EC 2003, *Financing Agreement - Urban Markets Development Programme*, European Commission, Brussels, Belgium.

Government of Zambia, (1997). *The markets act*. Available at: www.parliament.gov.zm/sites/default/files/documents/acts/Markets%20Act.pdf [Accessed 23 Feb. 2018].

Government of Zambia, (2015). *The urban and regional planning act*. Available at: www.parliament.gov.zm/sites/default/files/documents/acts/The%20Urban%20and%20Regional%20Planning%20%20Act%2C%202015.pdf [Accessed 23 Feb. 2018].

Hampwaye, G. (2008). *Decentralisation, local economic development and urban agriculture in Zambia*. Unpublished PhD, University of the Witwatersrand.

Hansen, K.T. (2004). Who rules the streets? The politics of vending space in Lusaka. In: K.T. Hansen, and M. Vaa, eds., *Reconsidering informality: Perspectives from urban Africa*. Uppsala: Nordic Africa Institute, pp. 62–80.

Hansen, K.T. (2010). Changing youth dynamics in Lusaka's informal economy in the context of economic liberalization. *African Studies Quarterly*, 11(2/3), pp. 13.

HLPE (High Level Panel of Experts), (2014). *Food losses and waste in the context of sustainable food systems*. Rome: High Level Panel of Experts on Food Security and Nutrition of the Committee on World Food Security.

Jongh, L. (2013). *The right to the city from a local to a global perspective: The case of street vendor and marketer organizations in urban areas in the Copperbelt, Zambia*. Unpublished Masters, Stockholm University.

Kachemba, N. (2017). Nakadoli market finally occupied. *Zambian Daily Mail*. Available at: www.daily-mail.co.zm/nakadoli-market-finally-occupied/ [Accessed 23 Feb. 2018].

Kalemba, A. (2011). *Assessing the value for money in public private partnerships infrastructure projects in Zambia: Case of Lusaka City Council*. Rotterdam: MSc Programme in Urban Management and Development, IHS, Erasmus University.

Kalero, R. (2017). Nakadoli traders owe KCC K1.5m levy. *Daily Nation*. Available at: www.dailynation.news/nakadoli-traders-owe-kcc-k1-5m-levy/ [Accessed 23 Feb. 2018].

Kalero, R. (2018). Market associations side-lined? *Daily Nation*. Available at: www.pressreader.com/zambia/daily-nation-newspaper/20180118/281676845323781 [Accessed 23 Feb. 2018].

Kamete, A.Y. (2013). Missing the point? Urban planning and the normalisation of 'pathological' spaces in southern Africa. *TransactionsInstitute of British Geographers*, 38, pp. 639–651.

UN-Habitat, (2009). *Zambia: Kitwe urban profile*. Nairobi: UN-Habitat.

KCC (Kitwe City Council), (Undated). *Council*. Available at: www.kcc.gov.zm/council-2/ [Accessed 23 Feb. 2018].

KCC (Kitwe City Council), (2012). *2012–2016 Strategic Plan*. Kitwe: KCC.

Larsen, M.M. (2009). *Institutional entrepreneurship in Zambian markets*. Masters, Copenhagen University.

Lungu, E. (2015). Speech at the opening of Mukuba Mall, Kitwe, 23 April 2015. Available: http://lusaka voice.com/2015/04/23/president-lungu-opens-mukuba-mall-in-kitwe [Accessed 23 Feb. 2018].

LusakaTimes, (2012). *Street vending needs to be controlled—Kambwili. LusakaTimes*. Available at: www.lusakatimes.com/2012/03/11/street-vending-controlledkambwili/ [Accessed 23 Feb. 2018].

LusakaTimes, (2013). Kitwe City Council slammed for failing to maintain the newly constructed Nakadoli market. *LusakaTimes*. Available at: www.lusakatimes.com/2013/02/08/kitwe-city-council-slammed-for-failing-to-maintain-the-newly-constructed-nakadoli-market/ [Accessed 23 Feb. 2018].

LusakaTimes, (2014). Kitwe Town Clark calls on vendors to return to markets. *LusakaTimes*. Available at: www.lusakatimes.com/2014/05/14/kitwe-town-clark-calls-vendors-return-markets/ [Accessed 23 Feb. 2018].

LusakaTimes, (2016). Government pledges to construct Chisokone B shelter that collapsed after a heavy downpour. *LusakaTimes*. Available at: www.lusakatimes.com/2016/12/18/government-pledges-construct-chisokone-b-shelter-collapsed-heavy-downpour/ [Accessed 23 Feb. 2018].

LusakaTimes, (2017a). Kitwe City Council Bans Street Vending as part of efforts to clean up the city. *LusakaTimes*. Available at: www.lusakatimes.com/2017/01/17/kitwe-city-council-bans-street-vending-central-business-district-part-efforts-clean-city/ [Accessed 23 Feb. 2018].

LusakaTimes, (2017b). Marketeers' body welcomes banning of street vendors in Mufulira. *LusakaTimes*. Available at: www.lusakatimes.com/2017/12/11/marketeers-body-welcomes-banning-street-vendors-mufulira/ [Accessed 23 Feb. 2018].

Mulenga, D. (2016). Zambia liberalises maize pricing effective 2018 marketing season. *African-Farming.com*. Available at: www.africanfarming.com/zambian-maize-prices-to-be-liberalised/ [Accessed 23 Feb. 2018].

Ng'anjo, N. (1994). *An evaluation of the law on street trading in Zambia (A case study of Lusaka)*. Bachelor of Laws Dissertation, University of Zambia. Available at: http://dspace.unza.zm:8080/xmlui/bitstream/handle/123456789/2844/NATHAN1.PDF?sequence=1 [Accessed 23 Feb. 2018].

Ntambi, M. (2017a). Chisokone market construction starts. *Zambian Daily Mail*. Available at: www.daily-mail.co.zm/chisokone-market-construction-starts/ [Accessed 23 Feb. 2018].

Ntambi, M. (2017b). Chisokone market drainage, sewer systems need overhauling—Kang'ombe. *Zambia Daily Mail*. Available at: www.daily-mail.co.zm/chisokone-market-drainage-sewer-systems-need-overhauling-kangombe/ [Accessed 23 Feb. 2018].

Scott, E.P. (1985). Lusaka's informal sector in national economic development. *The Journal of Developing Areas*, 20(1), pp. 71–100.

Songolo, S.K.C. (2017). Sustainable urban food policies through market infrastructure and logistics. In: J. Debru, S. Albert, N. Bricas, and D. Conaré, eds., *Urban food policies*. Proceedings of the international meeting on experiences in Africa, Latin America and Asia. Montpellier: United nations educational sceintific and cultural organization, pp. 27–35.

UKZambians, (2012). Kitwe city council u-turns on Chisokone. UKZambians. Available at: http://ukzambians.co.uk/home/2012/07/11/kitwe-city-council-u-turns-on-chisokone/ [Accessed 23 Feb. 2018].

10 Governance of food systems in Epworth, Zimbabwe

Easther Chigumira, Godfrey Tawodzera, Oliver Manjengwa, and Idah Mbengo

Introduction

Urban food systems in the global South are currently undergoing major transformation (Von Braun *et al.* 2008). However, despite the increasing body of work on urban food systems in this region, there is little work on how these food systems are governed, how governance structures impact on food production, distribution and consumption, and how this in turn effects food security and poverty reduction (Haysom 2014). This chapter examines the state of food system governance in Epworth, Zimbabwe, through a discussion on the food system actors, national policies and plans, legislation, economic instruments and extra-governmental forms of governance that shape and impact on the urban food system. Food system governance in this chapter is defined as the "interrelated and increasingly integrated system of formal and informal rules, rule-making systems, and networks at all levels of the food systems (from farm to fork)" (Bohle *et al.* 2009: 54).

Background and context

Urban governance is practiced through the planning, development and management of urban areas (Muchadenyika 2013). Ranger (2007) argues that the legislative frameworks, land use planning and measures of social control and repression that underpin the governance and management of urban areas in Zimbabwe today have their roots in the country's colonial history. As such, this section provides contextual background on the political economy, key legislation, government regulations, policies and planning practices that affect the food system. The first piece of urban governance legislation was the Town Management Ordinance of 1894, which gave legal effect to the establishment of Sanitary Boards. Successively, the Municipal Ordinance Act, promulgated in 1897, granted municipal status to the Sanitary Boards so that they could elect councils to manage urban areas.

Colonial governance was premised on the "bifurcation of the state" (Mamdani 1996) and a legal framework was provided for by the Land Apportionment Act of 1930. The Act institutionalized racial differences and created a class of white "citizens" and black "subjects" (Mamdani 1996). This segregationist approach was entrenched by the enactment of the Municipal Act of 1930, the Town Planning Act of 1933, the Urban Councils Act of 1973 and the Regional, Town and Country Planning Act of 1976, which was closely modelled on British planning law. These exclusionary racial policies resulted in the underdevelopment of African townships which became characterized by under-provision of services, high levels of poverty and informal economic activities (Mazingi and Kamidza 2011). The state was, however, averse to informality and ruthlessly destroyed both informal economic activities and housing (Mazingi and Kamidza 2011). This pathologization of informality was informed by a normative perception of an urban development that Kamete (2012) articulates as formal, ordered, healthy and aesthetically pleasing. This viewpoint continues to influence post-independence urban governance in Zimbabwe. In colonial Zimbabwe, urban governance was highly centralized (Ranger 2007) and land planning laws and policies were generally crafted to protect white settler commercial and trading interests.

Urban governance in post-colonial Zimbabwe

The socio-economic policies of both the colonial and independent governments have significantly impacted on the morphology of urban areas in Zimbabwe (Munzwa and Wellington 2010). This section discusses the governance of the country relating to four periods: 1980–1990 (Independence era); 1990–2000 (Economic Structural Adjustment Programme era); 2000–2009 (Fast Track Land Reform and Hyperinflation era); and post-2009 (Dollarization era).

1980–1990: Independence era

At independence in 1980, Zimbabwe inherited a highly centralized governance system and a dichotomous tripartite local government structure built on uneven development along racial lines (Jonga and Chirisa 2009). The post-colonial government introduced reforms to correct the racially based model of governance and facilitate devolution of responsibilities and power to local government. The Urban Councils Act of 1973, for example, was repealed and a decentralized form of local governance was adopted in line with the economic policy of "growth with equity". The Urban Councils Act of 1980 created a bimodal local governance structure consisting of rural and urban councils (Jonga and Chirisa 2009). The removal of repressive influx control legislation saw many Africans migrate to urban areas. This led to urban sprawl in the country and the development of secondary cities like Epworth. Despite legislative amendments aimed at reversing the colonial spatial and developmental inequalities, the state continued to be influenced by colonial policy and planning which regarded informality as abnormal and illegal (Kamete 2007). Central government continued to "interfere" and exercise control over local governance regardless of its devolution policy (Muchadenyika 2013). For example, Harare's first Master

Plan was prepared in the late 1980s (approved 1993) based on 1976 planning legislation (see Chapter 11).

1990–2000: Economic Structural Adjustment Programme (ESAP) era

The adoption of ESAP and subsequent market-oriented economic programmes increased urban poverty and modified the urban governance terrain. The introduction of executive mayors, provincial governors and provincial administrators entrenched central government in urban affairs. This period witnessed food-related riots and major civil unrest resulting from the negative impact of austerity measures underpinning ESAP: trade liberalization, devaluation and cessation of subsidies. Many companies closed, unemployment rose and informal economic activities increased (Tibaijuka 2005).

2000–2009: Fast Track Land Reform and Hyperinflation era

The government embarked on the controversial Fast Track Land Reform Programme (FTLRP) in 2000. Thereafter the country experienced a sharp economic decline, characterized by declining exports, reduced balance of payments, reduction in foreign currency earnings and low levels of foreign direct investment (Fox *et al.* 2007). By 2008, Zimbabwe was in a hyperinflationary environment owing to excessive printing of money (UNDP 2008). Poverty levels increased, food prices escalated and households suffered from severe food shortages. Politics became central to urban governance as government tried to control urban localities that it had lost to the opposition (Muchadenyika and Williams 2016). It is during this period that Zimbabwe witnessed *Operation Murambatsvina* where government demolished informal sector activities, destroying livelihoods and the informal food sector that had become critical in urban food supply (Potts 2006).

Post-2009: Dollarization era and the new constitution

Following the highly disputed 2008 elections, a "power-sharing" government was formed (Moyo 2011). In February 2009, this coalition government officially retired the Zimbabwean dollar in preference to using multiple foreign currencies, which reversed inflation and stimulated economic growth. Between 2010 and 2013 the economy grew by more than 10% per year. This growth, however, slowed down after 2013 due to poor agricultural harvests and decreased investment (ZEPARU 2014). Since 2013 many manufacturing companies have closed or scaled down operations (UNDP 2014). In 2015 alone, 4610 companies closed down and 55 443 jobs were lost (Government of Zimbabwe Budget Statement 2015). This has created space for further growth of the informal economy as a key urban livelihood strategy. Yet, the state has not changed its repressive policies toward informal practices in the economy. A landmark for this period that has had an influence on urban governance is the 2013 Constitution which recognizes local self-governance through Section 276 (2). This section entrenches the principle of devolution of powers from national to provincial and local governments (Mavedzenge 2012). In practice, however, power, control and decisions continue to be concentrated in national government, especially decisions on human resources, budgeting and revenue management (Muchadenyika and Williams 2016).

Research methods

In order to understand the nature of governance of urban spaces and its impact on the distribution and consumption of food (food systems) in Epworth, the Consuming Urban Poverty (CUP) project carried out two surveys in 2016: a food retail survey and a five-item value chain analysis. This primary data were supplemented by secondary data from the review of literature on urban governance and legislation in the country.

The food retail survey sought amongst other objectives to gain insight into the governance processes of food systems by (1) ascertaining food systems and retail operating processes and the retail environment in Epworth; (2) interrogating governance-related aspects such as store location, product sourcing, inspections and licensing and how this is engaged by traders, gatekeepers and officials; and (3) ascertaining the nature of support for traders and the wider socio-economic policy environment. Through purposive sampling three wards (1, 4 and 7) were selected for the retail survey in order to understand the spatial characteristics of urban food systems, and their governance thereof across high-, middle-, and low-income households. In total 297 retail services were surveyed.

The value chain analysis involved selecting five key food products consumed in Epworth: rice; mealie-meal (maize meal); vegetables (covo and rape); maputi (popcorn); and offal. Simple random sampling was utilized to select retailers trading in each of these products in each ward. Starting with these retailers, the foodstuffs were then tracked back to their sources. The purpose of tracing these products along the value chain was to determine the actors, institutions and policy instruments that influenced the governance of food systems in Epworth, as well as to identify the multiple packaging and pricing activities that occurred along the chain.

Governance of food systems in Epworth

Epworth's food system is embedded in broader national, regional and global food governance structures and policies, thus necessitating an understanding of these trends and policies. Global food governance is currently shaped by the mandates of Sustainable Development Goal (SDG) 2, which seeks to "end hunger, achieve food security and improved nutrition and promote sustainable agriculture". Zimbabwe's prioritization of the SDGs ranks SDG 2 third on its priority list for implementation. To achieve this, until 2030 government will focus primarily on improving food production in rural areas. This rationalization is based on the assumption that agriculture is the bedrock of Zimbabwe's economy. This production-centred approach, however, gives less attention to other value chain aspects that impinge on the urban food system.

Zimbabwe's constitution enshrines the right to food and food security through Sections 15 and 77. Section 15 mandates the state to encourage people to grow and store adequate food, secure adequate food reserves and promote adequate and proper nutrition (Government of Zimbabwe 2013a). Section 77 instructs the state to take reasonable measures to ensure food security (Government of Zimbabwe 2013a). Despite these provisions, the constitution still falls short as it puts emphasis on food production as a means to achieve food security. There is no explicit focus on other elements of food security (e.g. transport, distribution and pricing) which also shape food systems. Regardless, the Constitution provides a foundation for food governance that can encapsulate access to markets, urban food supply, and distribution to ensure food security of the urban poor.

Zimbabwe has never had a clearly articulated agricultural policy nor a specific urban food policy. Rather the food system has been governed through general policies and legal frameworks. In 2013, however, the government launched the National Food and Nutrition Security Policy, which aims to ensure adequate food and nutrition security in the country (Government of Zimbabwe 2013b). The policy reaffirms the "right to food" as detailed in the constitution, and reinforces the roles and responsibilities of communities in ensuring food and nutrition security (Government of Zimbabwe 2013b). An implementation framework for the policy is provided for by the National Nutrition Strategy (2014–2018) which envisions a Zimbabwe free from hunger and malnutrition. However, the policy and strategy focus mainly on the food availability dimension and inadequately speaks to other critical dimensions of urban food security: food access and food stability. In addition to the National Food and Nutrition Security Policy, the Zimbabwe Agenda for Sustainable Socio-Economic Transformation (ZimAsset), through its Food Security and Nutrition Cluster, provides a framework for policies and programmes in support of nutrition and food security for the country. Thus ZimAsset has shaped the governance of food systems such as Epworth's, through its drive for increased production for economic growth in the country.

Laws, policies and strategies

The management and governance of Epworth and its food system is mainly guided by the Regional, Town and Country Planning Act (RTCP) of 1976, which was amended in 1996 with no significant changes but that of colonial names and Urban Councils Act. These legislative instruments are steeped in colonial norms of land use planning, and only provide for mono-functionally zoned urban forms which do not take into account mixed land use patterns and the growing informality (Chirisa and Damba 2012). Furthermore, the RTCP Act requires that only landowners with title deeds and retailers with licenses be consulted in the planning process (Machakaire and Tapela 2016). Informal food sector players are therefore not consulted in the development of by-laws and the planning of market areas, thus militating against the ability of the urban poor to shape their governance structures.

Content analysis of the Epworth Local Board (ELB) by-laws and strategic plan (2016–2020) shows a general concentration on formalization and regulation of trade and land use planning. This is incongruent with the largely informal nature of Epworth. The by-laws enacted do little to facilitate street vending and urban food systems as they are crafted by the colonial vision of what an ideal city should look like. The current spatial distribution of vending and street vending activities across the seven wards in Epworth is thus viewed as disorderly, haphazard, and, in Kamete's (2012) terms, "abnormal".

Two national acts, the Public Health Act (2002 (1924)) and the Food and Food Standards Act (2001 (1971)) govern the vending of cooked foods in Zimbabwe. In Epworth, by-laws emerging from these acts are enforced. These laws do not allow vending of cooked foods and restrict what and where certain types of foods (e.g. vegetables) may be sold. Informal food vendors are therefore constantly evicted by municipal or national police. The ELB's involvement in the food system encompasses tight regulation and control through a diversity of economic, policy and legal instruments set by the national government. A good example is *Operation Murambatsvina* in 2005, where the government partnered with local authorities to enforce urban by-laws to stop "illegal" activities, including street vending and urban cultivation, in the process destroying vending stalls, flea markets and informal businesses.

The RTCP and Urban Councils Acts have not been re-aligned to the 2013 Constitution, which provides for individual rights to work and collective rights to public space and economic associations, including street vending. A SWOT analysis done for the ELB strategic plan (2017) highlights that almost 80% of the vendors were operating without licenses and in un-designated zones, while 40% of business operators did not have licenses. However, due to high unemployment and acute poverty, street vending, especially food trade, has emerged as a key livelihood strategy (Manjengwa *et al.* 2012). Study results also indicate that 74% Epworth residents access food through vendors/traders/hawkers. This is because vending allows for bulk-breaking and credit facilities, making food accessible to the poorest person (Ahmed *et al.* 2015). Yet current by-laws and regulations in Epworth continue to exclude and criminalize street trade, exacerbating poverty and subjecting the locality's food system to multiple forms of exploitative governance systems.

Extra-governmental forms of food governance in Epworth

Extra-governmental forms of food governance refer to the authority (institutions, regulatory bodies, cartels) and power relations that determine how financial, material and human resources are distributed along the food value chain (Matondi and Chikulo 2012). In studying African cities, market places have emerged as important spaces for urban food systems and key sites for urban food governance. Practices of governance in these spaces are exercised and contested by a variety of actors through various layers of relations involving various modes of power at different scales (Lindell 2008: 1880). In Zimbabwe generally, the informal retail sector plays an important role in providing the urban poor with access to food (Makaye and Munhande 2008). This sector, however, is also more susceptible to actors that set and enforce rules under which other actors in the value chain operate or benefit (Matondi and Chikulo 2012).

The CUP study showed that the governance of the food system in Epworth, outside of the state and council, is influenced by gatekeepers (*makoronyera*) and marshalls (*mahwindi*), a group of people who are akin to bus conductors, who collect fees from passengers on the commuter omnibus. Interviews with traders revealed that gatekeepers exercised their power by charging a protection fee to vendors who did not possess legal certification to operate. Their role entailed protecting vendors from council raids, facilitating licenses from the local board and supplying insider information on when raids would occur. The CUP retail survey found that 39% of the people in food retail did not have operating licenses. Such operators were most likely to pay protection fees for survival. According to study respondents, gatekeepers mostly targeted vendors operating along main roads or at shopping centres where large volumes of pedestrian and vehicle traffic result in competition for space.

Since the dollarization of the Zimbabwe economy in 2009, the country has seen the emergence of middlemen as key actors in the governance of the urban food system, resulting in food commodities moving through many hands along the value chain before reaching the consumer (Matondi and Chikulo 2012). Frayne *et al.* (2014) argue that in urban areas, food availability is seldom the major constraint, but rather lack of access to food for the urban poor. In Epworth, the *makoronyera* contribute to high price volatility of food especially fresh fruit and vegetables, thus curbing poor people's ability to access food. The reverse value chain analysis of five key products in Epworth revealed that there were four types of *makoronyera* in the food value chain: (1) those that operated in the trading of food; (2) those that linked directly with farmers; (3) those that were found at the food wholesale market; and (4) those that were involved in the auctioning of food products (Dhewa, Personnel Communication 2017). These *makoronyera*, operating as cartels, take advantage of economic or weather-related shocks to fix prices and often intimidate farmers to sell produce at low prices and retailers to buy at high prices. Such price fixing hits the poor hardest as vendors pass on price increases to the end consumer.

Results from the CUP retail survey indicate that women were dominating (52%) the retail trade in comparison to men (48%). Despite this dominance, however, female vendors indicated that they generally struggled to access fresh fruit and vegetable markets due to the presence of *makoronyera* and *mahwindi*. Some women indicated being forced to buy fruits and vegetables at high prices, which diminished their profit margins drastically. The *mahwindi* operating at Mbare Musika and open-air markets in Epworth have become a new type of actor in urban food governance. They charge a passage fee of between US$1 and US$2 for transporters to and from the market (Matondi and Chikulo 2012). Often, they act as brokers and source clients for the buyers and sellers for a fee, sometimes forcefully directing consumers to their preferred market stalls where prior financial arrangements would have been made for their benefit. This practice reduces competition among wholesalers as a "cartel" system kicks into play. As Figure 10.1 shows, both *makoronyera* and *mahwindi* are found along the key nodes of the maize and fresh fruit and vegetables value chain: (1) at the source of the product; (2) at the main markets such as Mbare; (3) en route to the retail markets; and (4) at the retail markets in Epworth. At each of these nodes, the price of the products increases substantially before reaching the consumer.

Thus, food provisioning and governance in Epworth is subject to various power relations and "strategies of influence" amongst a variety of actors (state, local board, gatekeepers) that ultimately determine by whom and by what price food is accessed. This ultimately influences food security, nutrition and stability.

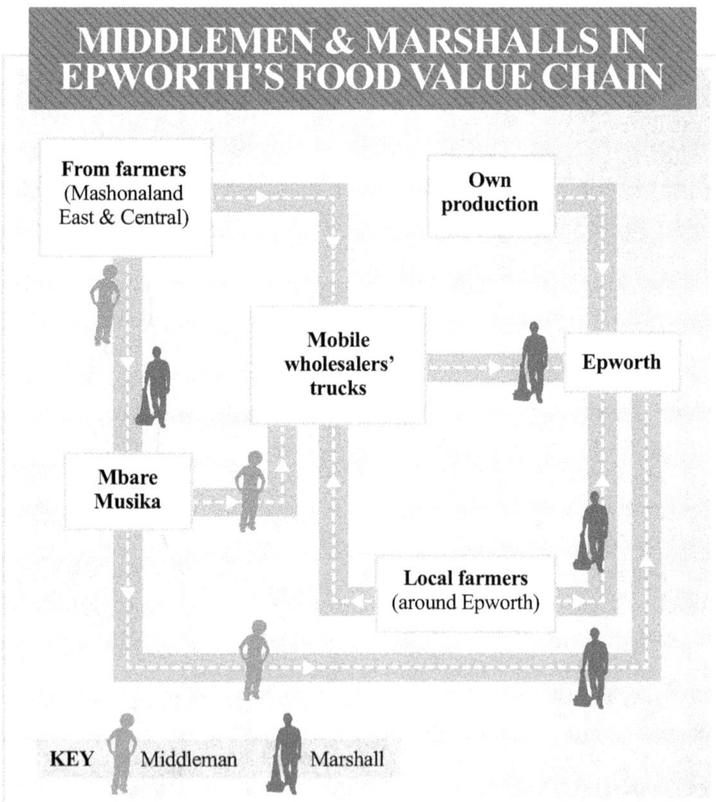

Figure 10.1 Cartel (middlemen and marshalls) operating on the food value chain for Epworth (Authors' own image)

Economic instruments and their effects on food governance in Epworth

A series of economic instruments introduced by national government, though not explicitly about food, have profoundly impacted the food system at both national and local levels. The Zimbabwe Agenda for Sustainable Development (ZimAsset), the country's economic blueprint, has four strategies for improving the food policy environment, which are to: (1) develop and review appropriate legislation, regulations guidelines and policies; (2) regulate food imports to promote local production; (3) monitor food imports and exports; and (4) invest in research, science and technology for agricultural development. Some of these strategies will be discussed to show how they impact on the food system, household food security and food governance.

The reverse value chain analysis of two of Epworth's five key food products tracked in this study, rice and maize meal, shows that these products were mostly imported from the Southern African region and from Asia (Figure 10.2a and 10.2b). This is because, since the Fast Track Land Reform Programme of 2000, Zimbabwe has become a net food importer. In addition, the dollarization of the economy has made local production more expensive, thus encouraging imports. The devaluation of the South African Rand, on the other hand, has made food produced in South Africa more attractive to the local market. Hence, most retail shops and informal markets find it profitable to sell imported foodstuffs.

Within this context and in line with the ZimAsset policy, the national government promulgated three legal statutes that affect the foods system: (1) Statutory Instrument 64 of 2016 (SI 64/2016) gazetted on 1 July 2016; (2) Statutory Instrument 20 of 2017 (SI 20/2017); and (3) introduction of surrogate currency – "bond notes" tradable only in Zimbabwe. SI 64/2016 listed 43 product categories that required an import permit and placed a quota on import quantities. The government argued that SI 64/2016 was aimed at protecting the country's local industry from cheap imports and to reduce the country's unsustainable import bill. Ultimately, this measure was envisaged to stimulate the diversification of local production, thereby creating local economic growth and employment.

The introduction of bond notes in 2016 was aimed at overcoming liquidity challenges that were manifesting in the dollarized economy. The "bond effect" has been twofold: first, it has created a black-market for the US dollar required for importation of goods, and second, it has increased the cost of transactions and therefore the price of goods. Although the RBZ decreed that the bond currency should trade on par with the US dollar, traders now demand a 10–50% premium on products bought in the surrogate currency (Nyamunda 2017). Thus, "bollarisation" (a term coined by Nyamunda (2017) to refer to the use of bond notes in Zimbabwe) has made food more expensive in Epworth through both the macroeconomic inflationary pressures and retail food price increases.

Epworth consumers complained of frequent price increases since the introduction of bond notes and promulgation of the two statutory instruments.

(a)

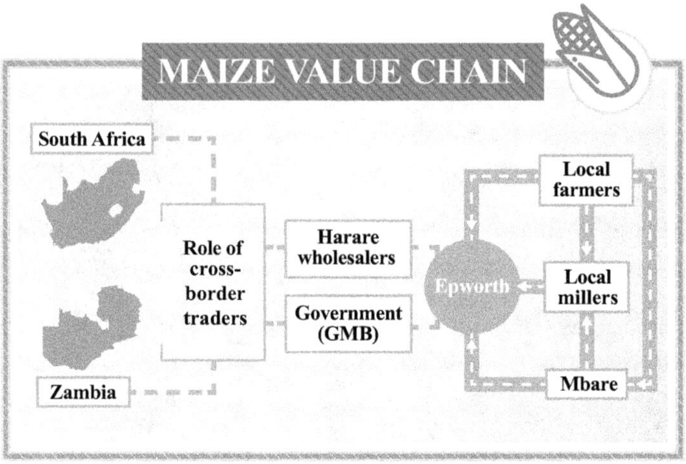

(b)

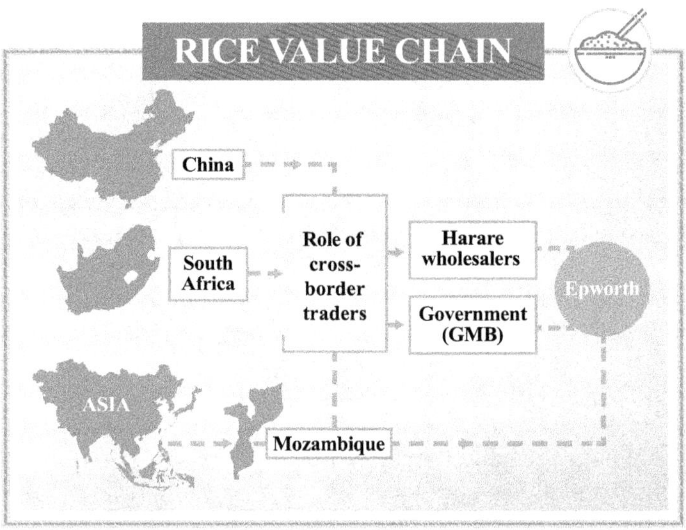

Figure 10.2 Value chain analyses of maize and rice

(Authors' own images)

Retailers complained of small profit margins as consumers were financially constrained, only purchasing food necessary for their survival. Most retailers also indicated that there was little paper money in circulation in Epworth. Without hard cash, most consumers have resorted to using debit cards. However, this study found that most traders do not possess point-of-sale machines to facilitate such transactions. Consumers wanting to transact using cards were forced to travel to Harare supermarkets for purchases, thus raising costs. Street traders and most food retailers were also losing considerable business as they did not have point-of-sale machines. It was also observed that a fifth type of middleman – the currency trader – has emerged in the food system, operating a money transfer system as households with bank accounts buy hard cash by transferring money through the Real Transfer Gross Settlement (RTGS) system to the middleman's bank account, usually at a premium of between 15% and 50%, further impoverishing already poor households.

SI 20 of 2017, effected in February 2017, imposed a 15% Value Added Tax (VAT) on rice, margarine, fish, potatoes and meat products such as offal, which are an important source of protein for the urban poor. The net effect was to increase prices of these products to unaffordable levels. Although the instrument was repealed seven days later due to consumer protest, most retailers in Epworth did not reduce their prices. Meat products are thus now out of reach for most Epworth residents. In Epworth, the price of meat (including offal) increased by between $1 and $2 per kilogram due to the introduction of VAT. Beef liver, which was $4.50 per kilogram in July 2016, was now priced between $5.50 and $6.50 per kilogram in February 2017. Such increases had devastating effects for the urban poor, most of whom stopped buying meat, thus impacting negatively on household food security and nutrition.

Conclusion

This chapter has shown the convergence of planning, economic and extra-governmental factors shaping Epworth's food system and governance. There is a disconnect between laws and planning paradigms on one hand and contemporary urbanisation dynamics on the other. Epworth's by-laws have not changed to take into account the fact that informal activities have become key to urban livelihoods in the country. Acts shaping Epworth's by-laws and policies have also not been re-aligned to the country's 2013 Constitution, which guarantees individual rights to work and collective rights to public space and economic associations (inclusive of street vending). As such, informal economic activities in the food system continue to be viewed as illegal and are ruthlessly suppressed. This, despite the fact that formality and informality symbiotically exist in Epworth. Exploitative middlemen and marshalls have also taken a centre stage in the governance structures of Epworth's food system, hiking prices and impoverishing consumers amidst the country's liquidity crunch. Any transition towards more inclusive governance of the food system thus needs to take into account these realities.

References

Ahmed, S., Simiyu, E., Gitliri, G., Sverdlik, A. and Mbaka, S. (2015). *Cooking up a storm, community-led mapping and advocacy with food vendors in Nairobi's informal settlements.* London: IIED.

Bohle, H.G., Etzold, B., Keck, M. and Sakdapolrak, P. (2009). Adaptive food governance. *IHDP Update*, 3(53), pp. 53–58.

Chirisa, I. and Dumba, S. (2012). Spatial planning, legislation and the historical and contemporary challenges in Zimbabwe: A conjectural approach. *Journal of African Studies and Development*, 4(1), pp. 1–13.

Dhewa, C. (2017). Personnel communication. Interview with head of eMkambo, Zimbabwe.

Fox, R., Chigumira, E. and Rowntree, K. (2007). On the fast track to land degradation? A case of the impact of the Fast Track Land Reform Programme in Kadoma District Zimbabwe. *Geography*, 92(3), pp. 212–224.

Frayne, B., McCordic, C. and Shilomboleni, H. (2014). Growing out of poverty: Does urban agriculture contribute to household food security in southern African cities? *Urban Forum*, 25(2), pp. 177–189.

Government of Zimbabwe, (2013a). *Constitution of Zimbabwe*. Harare: Government Printers.

Government of Zimbabwe, (2013b). *The food and nutrition security policy for Zimbabwe: Promoting food and nutrition security in Zimbabwe in the context of economic growth and development*. Harare: Government Printers.

Government of Zimbabwe, (2015). *2015 National budget statement*. Harare: Government Printers.

Haysom, G. (2014). *Food system governance for urban sustainability in the global south*. Unpublished PhD Thesis, Department of Environmental and Geographical Science, University of Cape Town: Cape Town.

Jonga, J. and Chirisa, I.(2009). Urban local governance in the crucible: Empirical overtones of central government meddling in local urban councils affairs in Zimbabwe. *Theoretical and Empirical Research in Urban Management*, 3(12), pp. 166–182.

Kamete, A.Y. (2012). Interrogating planning is power in an African city: Time for reorientation?. *Planning Theory*, 11, pp. 66–88.

Kamete, A.Y. (2007). Cold-hearted, negligent and spineless? Planning, planners and the rejection of filth in urban Zimbabwe. *International Planning Studies*, 12(2), pp. 153–171.

Lindell, I. (2008). The multiple sites of urban governance: Insights from an African city. *Urban Studies*, 45, 9, pp. 1879–1901.

Machakaire, D.G. and Tapela, T.N. (2016). Urban development planning practices and responses to rapid urbanisation in Harare: A case of dirty lenses, blunt frameworks/instruments or both? *SACSC conference, Working Paper*. Johannesburg: Southern African Coities Studies (SACSC)

Makaye, P. and Munhande, C. (2008). The contribution of informal food trading to urban food security in Zimbabwe. *Journal of Sustainable Development in Africa*, 10(2), pp. 312–332.

Mamdani, M. (1996). Citizen and subject: Contemporary Africa and the legacy of late colonialism. Princeton, NJ: Princeton University Press.

Manjengwa, J., Chimhowu, A.K., Chingarande, I., Masunda, W. and Kujinga, K. (2012). Household consumption and poverty in Zimbabwe. In: J. Manjengwa, S. Feresu, and A. Chimhowu, eds., *Understanding poverty, promoting wellbeing and sustainable development: A sample survey of 16 districts of Zimbabwe*. Harare: IES, pp. 29–43.

Matondi, P. and Chikulo, S. (2012). *Governance over fruit and fresh vegetables in Zimbabwe: Market linkages and value chain study*. Harare: Ruzivo Trust.

Mavedzenge, J. (2012). *Thematic analysis of Zimbabwe's proposed draft constitution*. Harare: Freedom House.

Mazingi, L. and Kamidza, R. (2011). Inequality in Zimbabwe. In: H. Jauch and D. Muchena, eds., *Tearing us apart: Inequalities in South Africa*. Johannesburg: Open Society Initiative for South Africa, pp. 322–383.

Moyo, S. (2011). Land concentration and accumulation after redistributive reform in post settler Zimbabwe. *Review of African Political Economy*, 38(128), pp. 257–276.

Muchandenyika, D. (2013). The crisis of local government. In: I. Chirisa, ed., *Contemporary rural and urban planning issues in Zimbabwe: Implications for policy and planning*. Palo Alto: Academica Pr Llc, pp. 229–248.

Muchadenyika, D. and Williams, J.J. (2016). Social change: Urban governance and Urbanisation in Zimbabwe. *Urban Forum*, 27(3), pp. 253–274.

Munzwa, K. and Wellington, J. (2010). Urban development in Zimbabwe: A human settlement perspective. *Theoretical and Empirical Researches in Urban Management*, 5(14), pp. 120–146.

Nyamunda, T. (2017). Economic prospects of SI64 in a bond notes and smuggling economy. *NewsDay Zimbabwe*. Available at: https://www.newsday.co.zw/2017/05/economic-prospects-si-64-bond-notes-smuggling-economy/

Potts, D. (2006). 'Restoring Order'? Operation Murambatsvina and the urban crisis in Zimbabwe. *Journal of Southern African Studies*, 32(2), pp. 273–291

Ranger, T.O. (2007). City versus state in Zimbabwe: Colonial antecedents of the current crisis. *Journal of Eastern African Studies*, 1(2), pp. 161–192.

Tibaijuka, A. (2005). Report of the fact-finding mission to assess the scope and impact of 'Operation Murambatsvina' in Zimbabwe. UN-Habitat Special Envoy on Human Settlements, 18 July 2005. New York: United Nations.

UNDP, (2008). *Zimbabwe human development report*. Harare: UNDP.

UNDP, (2014). *Zimbabwe human development report*. Harare: UNDP.

Von Braun, J., Ahmed, A., Asenso-Okyere, K., Fan, S., Gulati, A., Hoddinott, J. and van Rheenen, T. (2008). *High food prices: The what, who and how of proposed policy actions*. Washington, DC: International Food Policy Research Institute.

ZEPARU, (2014). *Zimbabwe economic review*. Presented at the Zimbabwe Economic Policy Analysis and Research Unit, Victoria Falls.

11 Urban food production in Harare, Zimbabwe

Percy Toriro

Introduction

As the preceding chapters have illustrated, there has been little policy attention paid to the governance of the urban food systems in African cities. The only area where there has been significant policy and NGO interest has been urban agriculture. However, despite the current efforts to enable urban food production, the treatment by the State even of this apparently well-supported food system intervention has been inconsistent and, at times, contradictory. Given the importance of urban agriculture as the State's primary engagement in urban food system governance, this chapter considers the trajectories of urban agriculture policy in Harare (Zimbabwe).

This chapter provides an account of the development of statutory planning regulations, policies, and different stakeholder roles and perspectives in Harare with regards to urban agriculture. It considers the history of these developments in the context of the specific vision of urbanity that has driven them and of the influence of international development discourses in shaping local practice. It covers the period from 1890 when Harare was founded, and when food production was first recorded in the city, to the present.

The chapter builds on Mbiba's (1995) early work on urban agriculture in Zimbabwe, bringing focus to the legislative, policy, and spatial frameworks that impact urban food production. While Mbiba (1995) discussed regulatory challenges only briefly, this chapter seeks to demonstrate that without addressing the legislative, policy, and planning inconsistencies, the potential for meaningfully and sustainably addressing food production in the city is lost.

While the uptake of urban agriculture varies significantly between low-income areas in different African cities (Crush *et al.* 2011), the prevalence of urban agriculture is high in Harare. The 2008 African Food Security Urban Network (AFSUN) survey in low-income neighbourhoods of the city found that as many as 60% of sampled households were producing food (Tawodzera *et al.* 2012). This could be accounted for in part by the sustained economic decline in Zimbabwe over the last decade and half (Toriro 2009; Tawodzera *et al.* 2012). The wider AFSUN survey, conducted in low-income areas of 11 southern African cities, found that food-insecure households were far more likely to grow food than food-secure ones (Crush and Frayne 2010b: 41).

This chapter argues that although food production has always taken place in Harare, its management has not been smooth as it has been subject to shifting positions as well as limited acceptance by planning professionals. Although urban food production is accepted in the Harare Master Plan as well as other national policy pronouncements, the failure by professionals to cascade higher level policy to the local level creates a potential challenge in the governance as food production remains exposed to the discretion of officials whose values may be out of alignment with the lived reality of urban residents.

The next section of the chapter looks at the shifting policy responses to food production in Harare over time. This is followed by an analysis of the Harare Master Plan looking at specific clauses that have to do with urban food production as well as the current application of the plan and the attendant policies. This section shows that although there is now less harassment of producers, there has been a mixed response and interpretation of the plans and policies over time. Evidence will be drawn from interviews conducted by the author with planning professionals and from published reports.

Shifting food production and policy responses in Harare

Early acceptance of urban food production in the city

Urban food production was an integral part of early spatial planning in Harare. In one of the earliest suburbs to be established south of Harare, Waterfalls, low-income white (or non-indigenous) residents were allocated larger horticultural allotments on which they were allowed not only to set up their residential units, but also to grow food for their own consumption and sale (Zinyama 1993; City of Harare 1996). In the east of Harare another suburb for domestic workers, with two sections called Mabvuku and Tafara, was also established in the late 1960s with a large area of land demarcated for agricultural allotments and managed by the area municipal superintendent. The facilitation of food production was a means to support poor families' livelihoods (City of Harare 1973; Zinyama 1993). There is further evidence of formal agricultural production in a few other areas such as Tynwald and Marlborough suburbs and Epworth. In Epworth food production was also tolerated and encouraged by local officials on the plots that were allocated to the initial settlers. Much of this production has, however, been significantly constricted due to the continued unplanned residential growth in this area over the years.

Heavy-handedness and inconsistency of municipal officials

Conflict over food production within Harare only developed after independence in 1980. Prior to this there was abundant vacant urban land, and rural–urban migration was limited due to "influx control" imposed by the colonial government (Potts 2010). However, after independence all discriminatory laws that served to control permanent settlement of Africans in urban areas were removed (Potts and Mutambirwa 1990). As many Africans pursued real and perceived opportunities in Harare, demand for resources, including land, increased. Previously undeveloped public lands that had been available for food production were built on and residents therefore made use of the remaining more marginal and ecologically fragile land for food production. Some of this cultivation was seen by authorities as violating environmental regulations, the most important one being a restriction on cultivation within 30m of stream banks and rivers (Mbiba 1995). City authorities would periodically send municipal police to destroy thriving maize crops, some nearing maturity, thereby attracting public anger.

The practice of destroying crops was never consistent. In some years inter-agency teams (composed of officers from different municipal departments, national government, and the government Natural Resource Board) would destroy crops, yet in other years they would turn a blind eye (Mbiba 1995). There was also inconsistency between city officials and policy makers. Perhaps the most striking example of this inconsistency, as identified in an interview with the current head of Town Planning for Harare, was in the early 1990s when the head of town planning in the city ordered the destruction of "illegally grown" maize. Unknown to the head of planning, the then Mayor of Harare, Mr. Solomon Tawengwa, had supported the same farmers by providing inputs such as seed and fertilizers to the farmers. After the Mayor was informed about the destruction of the crops, the planning official was immediately dismissed from his position. It took the court of law for the official to be re-instated on the justification that he was only implementing environmental and health regulations. But more importantly the fact that one arm of the municipality was promoting what the other was fighting against demonstrates the inconsistencies in the governance of food production over time (Mbiba 1995).

The economic structural adjustment programme, economic decline, and urban food production

In the 1990s the government of Zimbabwe introduced an economic structural adjustment programme (ESAP) which impacted on food production in Harare. Whilst the intention of the programme was to open up the economy and attract foreign direct investment, the result was catastrophic for the economy (Potts 1998). Local industry, which had thrived under protectionist policies, buckled under increased competition from cheap imports. This resulted in many company closures leading to unemployment. The ESAP was a major turning point in local food production in Harare. Studies conducted during the period indicate a notable rise in the land under cultivation, such that ENDA-Zimbabwe (1996) found that the amount of space taken by agriculture in Harare grew from 4822 hectares in 1990 to 9828 hectares in 1994 (ENDA-Zimbabwe 1996; Toriro 2009). Chaipa (2001) observed that the total land under cultivation in the city increased by over 90% in the early 1990s alone. This growth in own food production was driven by increasing poverty and the need to seek alternative livelihoods.

Seasonal variability: the 1992 and 2002 drought

Over the period in which the ESAP was indirectly leading to increased food production within Harare, the country experienced two severe droughts, in 1992 and in 2002. Agriculture failed to supply the staple food crop, maize, and the country had to import grain to feed the people. Whilst crops in many agro-ecological zones completely failed, in Harare crops were in a somewhat better condition. City officials found themselves in a dilemma: how to morally justify destroying crops that are doing well in a period when everywhere else there were gloomy stories of wilting crops? Based on information obtained in an interview with a City Planning official, it appears that authorities quietly decided against destroying crops in the previously condemned "illegal" areas. This became a turning point that probably influenced the 2002 Nyanga Declaration on Urban and Peri-Urban Agriculture, a public policy position taken by the association of urban local governments in Zimbabwe at their annual conference in 2002.

NGO interventions

The late 1990s and the 2000s saw an increased interest in urban food production by NGOs that had previously concentrated their attention on agriculture in rural Zimbabwe (Potts 2008). An urban agriculture stakeholders' forum, supported by NGOs, that lobbied for the Nyanga Declaration was formed in Harare. This brought together urban food producers, government extension workers, municipal planning and social services officials, NGOs supporting urban food, community leaders, and academic institutions. In an interview with a Municipal Development Partnership for Eastern and Southern Africa (MDP-ESA) official it was found out that two NGOs (the Zimbabwean Environmental Lawyers Association (ZELA) and MDP-ESA) that were working to lobby for the creation of an enabling policy and legislative environment for urban agriculture commissioned a study to analyze Zimbabwe's laws to see if they inhibited urban agriculture. ZELA also produced guidelines to assist anyone wanting to start urban food production. Other NGOs supported herbal gardens that focused on producing highly nutritious food crops with medicinal properties perceived as effective in the fight against HIV/AIDS in Harare and other urban areas.

Policy improvement: the Harare and Nyanga declarations

There have been some significant policy shifts in the authorities' position on urban food production in Harare, starting from the early 1990s. This has largely been driven by national policy changes rather than municipal plans. Save for the pronouncements found in the Harare Master Plan of 1993 (discussed in a later section), which lays out the overall planning policy guide for the city and sets out general development parameters and policies, there are no other documented positions on urban agriculture for Harare. In 2002 all 31 Zimbabwean urban local governments met in the resort town of Nyanga for the annual conference of the Urban Councils Association of Zimbabwe (UCAZ). A discussion on urban food production led to an agreement that all urban areas in Zimbabwe needed to acknowledge the role of urban agriculture in promoting urban food security and livelihoods. An interview with ZELA revealed that this decision was largely led by the NGO lobby on behalf of the urban producers. They further agreed that they would, from that date onwards, work to facilitate local food production in all their urban areas. This became known as the Nyanga Declaration on Urban and Peri-Urban Agriculture 2002 (Box 11.1).

Box 11.1 Nyanga Declaration on Urban and Peri-Urban Agriculture in Zimbabwe, 7 July 2002, Nyanga, Zimbabwe

We, the delegates to the Urban Councils Association of Zimbabwe's 61st Annual Conference, comprising of the Minister of Local Government, Public Works and National Housing and his Deputy, the Minister of State for Informal Sector, Executive Mayors and Chairpersons, Town Clerks, Chairpersons of Council Committees, Councilors, Heads of Departments, representatives from the Ministry of Local Government, Public Works and National Housing, International, Regional and Local Non-Governmental Organizations, meeting at Montclair Hotel, Nyanga, from 4–7 June 2002, acknowledging that Urban and Peri-urban Agriculture contribute to urban food security, poverty reduction, local economic development and sustainable urban development

Are urging

Local authorities to promote Urban and Peri-urban Agriculture in their cities, develop appropriate incentives and other policies necessary for its growth, mainstream urban food security within their operations and promote the collection and dissemination of information on Urban and Peri-urban Agriculture activities in their territorial planning areas,

The Government to include Urban Agriculture in its programmes to alleviate poverty and economic empowerment, food security, promotion of local economic development and environmental and health improvement,

Non-Governmental Organizations and Donors to support financially and materially urban and per-urban agriculture projects for the benefit of the urban poor,

The private sector to invest in high value intensive urban and peri-urban agro-industries in order to create employment opportunities and promote local economic development.

Are encouraging

Urban local authorities to recognize the significance of the contribution of Urban and Peri-urban Agriculture to social development approaches, generation of jobs and income, self-esteem, environmental improvement and urban food security and to add them to their key development goals.

Are re-affirming

Our commitment to improve urban management through the promotion of peri-urban agriculture in our cities so as to enhance urban food security, address urban poverty, improve urban environmental and health management as well as to protect the urban and peri-urban bio-diversity.

7 June 2002 Nyanga, Zimbabwe

(www.cityfarmer.org/zimbaDeclare.html)

The Nyanga Declaration was a major milestone in changing the urban food production policy framework across Zimbabwe, providing clarity and momentum for subsequent complimentary actions. The following year a regional workshop on urban food production was organized by the government of Zimbabwe and funded by the Food and Agriculture Organization (FAO), Resource Centre for Urban Agriculture and Food Security (RUAF), United Nations Development Programme (UNDP), United Nations Children's Education Fund (UNICEF), International Development Research Centre (IDRC), and Food Agriculture and Natural Resources Policy Analysis Network (FARNPAN). The funding organizations were all working in the region to promote urban food security through urban food production. The participants were drawn from the governments in Eastern and Southern Africa, development partners, and some local governments. Following discussions at this workshop, participants declared a commitment not only to create a conducive framework for urban food production, but also to support it with resources. This commitment, signed by five governments, became the Harare Declaration on Urban and Peri-Urban Agriculture of 2003 (Box 11.2).

Box 11.2 The Harare Declaration on Urban and Peri-Urban Agriculture in Eastern and Southern Africa 2003

Preamble

We, the Ministers responsible for Local Governments from Kenya, Malawi, Swaziland, Tanzania and Zimbabwe, at our meeting in Harare on Urban and Peri-urban Agriculture (UPA) in Eastern and Southern Africa organized by the Ministry of Local Government, Public Works and National Housing of the Government of Zimbabwe and the Municipal Development Partnership for Eastern and Southern Africa, in collaboration with UNDP, UNICEF, FAO-SAFR, FANRPAN, RUAF and IDRC held on 28 and 29 August, 2003;

Acknowledging

The presence of local government practitioners and representatives of non-governmental organizations and community-based organizations;

Acknowledging further that

- UPA is a widely practiced activity in and around towns and cities within the region on parcels of land with alternative competing uses;
- UPA has generally been practiced informally without appropriate policy, legislative and institutional frameworks;
- UPA plays, and will continue to play, a significant role in promoting food security, employment creation and income generation, health and nutrition and improving the economies of urban areas;
- Some governments in the region have made significant progress in incorporating UPA in their urban development plans, and that others are now beginning to rise to the challenge,

Recognizing

The existence and increasing practice of UPA and also noting the many challenges that it faces, including:

- Absence, inadequacy and or inconsistency in the policies, legislation and institutional arrangements for regulating the sector
- Limited availability of and access to resources
- Limited research, documentation and information-sharing nationally and regionally
- The need for environmental sustainability

Accepting

That the foregoing challenges require immediate and prudent reform of policies, legislative and institutional arrangements in order to effectively integrate UPA into our urban economies,

We therefore

Call for the promotion of a shared vision of UPA that takes into account the specific needs and conditions in the region, and accordingly commit ourselves to developing policies and appropriate instruments that will create an enabling environment for integrating UPA into our urban economies.

Thus done at Harare on 29th Day of August, 2003

Signed: ____________________ Signed: __________________
Millicah W. Thairu Honourable Henry Midiani
For the Government of Kenya For the Government of Malawi
Signed: ____________________ Signed: ____________________
Honourable Albert Shabangu Honourable Mizengo Pinda
For the Government of Swaziland For the Government of Tanzania
Signed: ___________________________
Honourable Ignatius Chombo
For the Government of Zimbabwe

(Copy provide by the MDP-ESA)

The Nyanga and Harare Declarations softened the attitude of local government officials to urban food production. Since these two declarations, a few developments took place to demonstrate the improving urban food production environment. Firstly, the national government appointed and subsequently increased the number of agriculture extension workers in Harare and its satellite towns including Epworth (Interview with Ministry of Agriculture official). Secondly, government created and allocated one-hectare plots to beneficiaries in Harare, Epworth, and Chitungwiza for food production. According to a provincial government official, the people allocated these plots were selected through NGOs working on urban agriculture and through political parties. According to an interview with an official from the Minister of State for Harare Province's office, more than 1500 beneficiaries were allocated agriculture plots measuring one-hectare each.

The 2005 Operation Murambatsvina and major shift on urban agriculture

Whilst the above declarations helped remove conflict in urban agriculture governance, a major government action targeting urban informality in 2005 reversed most of the gains that had been made. This was known as Operation Murambatsvina, which is a Shona expression meaning to sweep away filth or dirt. Epworth was spared much of the destruction since it was considered an informal settlement being upgraded. This government-driven programme targeted all buildings, uses, and livelihood activities deemed illegal in all urban areas of the country, though most visibly in Harare. The impact of the operation on urban livelihoods for the urban poor was devastating (Potts 2010; Tawodzera 2013). More than 700 000 households were reportedly affected across the country's urban areas (Tibaijuka 2005). For food production this resulted in crops, mushroom growing structures, and food processing spaces being destroyed. Some fields lay fallow for a season or two because the farmers had been displaced from the area. However, by 2012, significant elements of informality had "bounced back" (Tawodzera *et al.* 2012: 31). Interviews with City Planning officials confirmed that for most of the disrupted urban agriculture activities, it is now "back to business as usual."

Operation Murambatsvina demonstrated the uncertainty surrounding the governance of food production in general, and informal food production in particular. The next section looks at statutory plans and planning legislation and how they have affected food production as important tools of governance.

Spatial and statutory planning: the Harare Master Plan (1993) and urban agriculture in Harare

This section focuses on the historical development of urban food production in the spatial and statutory plans of the city. The principal land use planning policy document for any city in Zimbabwe is the master plan in terms of the Regional Town and Country Planning Act of 1976: Chapter 29:12 (GoZ 1996). Harare's first Master Plan was prepared in terms of the Regional Town and Country Planning Act of 1976 in the late 1980s and approved in 1993 (City of Harare 1993). According to an interview with a central government planning department official, although it was supposed to be reviewed every ten years, there has yet to be a revision. According to an interview with a City Planning official, prior to the Harare Master Plan the city's planning was managed through Town Planning Schemes and local plans (following the creation of the Master Plan in 1993, new town planning schemes were termed "local plans"). The Harare Master Plan is a combination master plan, as it includes more than one administrative authority or district. It covers Greater Harare and includes the satellite urban areas of Epworth, Norton, Ruwa, Mazowe, and Zvimba. This is a large area covering up to a 60km radius from Harare city centre. The Master Plan has policies and strategies to guide development of the city. There are also specific provisions that indicate the city's position on food production. Policy numbers

(v) and (xii) are specifically worthy of note here. The Master Plan provisions are closely aligned with the wider initiatives and agreements on urban agriculture discussed in the previous section.

Policy (v) of the Harare Master Plan of 1993 states as follows: "*Insist on intensive productive use of agricultural land adjacent to/close to developed areas for perishables in order to provide for local consumption and/or export.*" According to a City Planning official, the intention here is an attempt to encourage the production of food within the city's hinterland to minimize the distance travelled by perishable food products. However, this clause in the Master Plan has been contradicted by official planning practice and practitioners' views, as will be demonstrated later.

Policy (xii) of the Harare Master Plan of 1993 states: "*Identify and zone land suitable for residential agriculture allotments.*" This provision, as well as Policy (v), provides a guide to how all other statutory and land use plans in the Harare city-region are supposed to be prepared. Although the 1993 Master Plan has clauses on food production, there has not been consistency in ensuring the policy positions are implemented. A small number of local plans, such as those for Hatfield/Waterfalls and Tynwald (both areas with long histories of space allocated for production), have areas zoned for food production. However, the clear majority of those plans prepared after the Harare Master Plan have no provisions for local food production. City officials have therefore deliberately, or out of omission, worked against food production. This general absence of food production in the local plans indicates a disconnect between planning intention and practice. Whilst the land use policy identified the need for food production, urban planners do not appear to see the need for food production in urban areas. Planners are supposed to lead in giving practical effect to the realization of the policy position, but they have not. As a result, most plans remain silent about food production. There are also contradictions amongst different planning officials, which were exposed in interviews:

> The plans have never been a problem; it is the attitude of the planners. I have heard it said many times that planning does not provide for urban food production but that is a lie. If it was a fact, why would our Master Plan provide for agricultural allotments and the maintaining of agricultural and horticultural zoning in areas with soils deemed fertile and productive? There are still highly productive areas which are recognized in plans in Waterfalls and Tynwald; these are official approved city plans. This tells you whoever is saying food production is not allowed in the city is expressing their own value and not the technical or official position.
>
> Interview with City Planning Official 1 (April 2017)

In Harare, as in Zimbabwe more broadly, food production is generally provided for in rural and commercial farming areas where it is the primary land use activity. Land in cities is subject to severe competition. And the principal 5 land use zones are residential, commercial, industrial,

> recreational, institutional and public utilities. You cannot preoccupy yourself with food in a city. We are there to deliver services, and food certainly is not one of the stated services.
>
> Interview with City Planning Official 2 (April 2017)

Out of Harare's more than 40 local plans, only four were found to have expressly planned food production spaces. The inclusion of urban food production within the Master Plan has little applied meaning, given its absence in the local plans. This absence can be largely attributed to the officers working in the planning system who are supposed to prepare local plans or supervise plans prepared by consultants. In both instances they determine the content of the plans. The planners that do not support urban agriculture seem to think it is "not proper" to allow a "non-urban" land use in town. It is further driven by a notion that food production in towns lowers "standards."

Resistance by planners to urban food production is also often informed by incorrect interpretation of key legislation. More than half of the public officials who were interviewed identified the law as constraining them from facilitating food production. There are four main laws that are usually cited as informing the regulation of urban food production in Zimbabwe, namely: The Regional Town and Country Planning Act (1996), The Urban Councils Act (2002), The Environmental Management Act (2003), and The Public Health Act (2001).

Because of this perpetually stated position that Zimbabwe's laws inhibit food production in urban areas, a study was commissioned in 2004 to verify that position. The main findings of that study were as follows:

1 That urban food production was not a prohibited activity;
2 That the law only provided for the 'regulation' of food production;
3 That the current legislative regime, whilst generally perceived as inhibiting urban agriculture, was only so where:

    a Urban food production violated ecological concerns;
    b Urban food production violated sound town planning;
    c Urban food production violated public health concerns.

(Mushamba and Makonese 2004)

This study theoretically provided clarity for local authorities; however, because it was produced by two NGOs during a period of heightened distrust between government and NGOs, its findings did not have significant impact on local authority perceptions or actions, and the myth that the law is problematic persists in public offices in Harare and elsewhere. Whilst the 2013 Constitution has recognized the right to food, it leaves the lower-tier laws to provide the detail, and these details are often poorly understood by planners.

Planners attitudes have been one of the major obstacles to the establishment of a clear food production framework for Harare and the secondary cities. Some Zimbabwean scholars attribute this to the planners' modernist view (Kamete 2013) and the interference of politics in planning in Zimbabwe (Muchadenyika and Williams 2017). This was also confirmed in interviews with practicing planners, as indicated earlier. Both the planners' attitudes and the role of politics are important in Zimbabwe's urban food production. The attitudes have affected both the growth and the sustainability of food production.

Conclusion

It is clear urban food production has always taken place in Harare, but the reaction of authorities has varied from the early accommodation of colonial times, targeting poor urban whites and African migrant workers, to the heavy-handedness of officials after political independence when the demand for land increased. Although the principal land use policy document of the city, the Master Plan, recognizes the need for urban food production, there has not been consistent implementation of the policy in the local plans. This position appears to be driven by many of the city planners, who appear unconvinced that urban food production should be provided for in the city. Therefore, whilst there may be overall acceptance of urban agriculture in the city, this position is not always reflected in the detailed plans. Whilst the national policy positions have meant that producers should not be harassed, the risk is that the position has not been formalized at the local level in planning. Although few Harare local plans have incorporated urban agriculture as a land use zone, outside of Operation Murambatsvina there are no documented cases of destruction of produce, even for those grown in "fragile ecological areas." In addition to observational confirmation of this, it was also confirmed in an interview with the current head of planning for the city of Harare. Both politicians and administrative officials seem to just turn a blind eye to the practice of urban agriculture even when it contravened official land use regulations. On the ground there are several NGOs working with communities on urban agriculture with the knowledge and/or approval of city policy makers and officials. However, this should not give comfort to stakeholders. Given past treatment of unregulated activities, authorities may yet take contradictory action. For as long as this reprieve is discretionary, and not backed by laws and statutory plans, the management of food production remains uncertain and producers are still vulnerable.

References

City of Harare, (1993). *Harare combination master plan: An approved statutory plan prepared in terms of the regional town and country planning act*. Harare: City of Harare.

City of Harare, (1996). *Hatfield waterfalls local development plan number 26: An approved statutory plan*. Harare: City of Harare.

City of Harare, (1973). *Cleveland local subject plan: An approved statutory plan for city of Harare*. Harare: City of Harare.

City Planning Official 1, (2017). *Interview with city planning official on their understanding of the Harare Master Plan position on urban food production*.

City Planning Official 2, (2017). *Interview with city planning official on their understanding of the Harare Master Plan position on urban food production*.

Chaipa, I. (2001). The urban planning Dilemma in Harare, Zimbabwe. *The UA Magazine, RUAF, Leusden*, 4, pp. 17–18.

Chikowore, E. (1993). Harare: Past present and future. In: L. Zinyama, D. Tevera, and S. Cummin, eds., *Harare, the growth and problems of the city*. Harare: University of Zimbabwe Publications.

Crush, J., Horvoka, A. and Tevera, D. (2011). Food insecurity in Southern African cities: The place of urban agriculture. *Progress in Development Studies*, 11(4), pp. 285–305.

Crush, J. and Frayne, B. (2010a). *The invisible crisis: Urban food insecurity in Southern African Cities*. Cape Town: AFSUN.

Crush J. and Frayne, B. (2010b). *The state of southern food insecurity in Southern Africa*. Cape Town: AFSUN, Unity Press

ENDA Zimbabwe, (1996). *Urban agriculture in Harare: Results and recommendations of a household survey conducted in Harare*. Zimbabwe: ENDA (unpublished).

GoZ, (1996). *The regional town and country planning act: Chapter 29:12. Revised Edition*, Harare: Government of Zimbabwe, Government Printers.

Kamete, A.Y. (2013). Missing the point? Urban planning and the normalisation of 'pathological' spaces in southern Africa. *Transactions of the Institute of British Geographers*, 38(4), pp. 639–651.

Mbiba, B. (1995). *Urban agriculture in Zimbabwe: Implications for urban management and poverty*. Avebury, Aldershot.

Muchadenyika, D. and Williams, J.J. (2017). Politics and the practice of planning: The case of Zimbabwean cities. *Cities*, 63, pp. 33–40.

Mushamba, S. and Makonese, M. (2004). *A review and analysis of the policy and legislative framework for urban agriculture in Zimbabwe*. Harare: Municipal Development Partnership and Zimbabwe Environmental Lawyers Association.

Potts, D. (1998). The impact of structural adjustment on welfare and livelihoods: An assessment by people in Harare. In: N. Nelson and S. Jones, eds., *African urban poverty*. London: Intermediate Technology Publications, pp. 36–59.

Potts, D. (2008). Displacement and livelihoods: The longer-term impacts of Operation Murambatsvina. In: M. Vambe, ed., *Zimbabwe: The hidden dimensions of Operation Murambatsvina*. Harare: Weaver Press, pp. 53–64.

Potts, D. (2010). *Circular migration in Zimbabwe & contemporary sub-Saharan Africa*. London: Boydell and Brewer.

Potts, D. and Mutambirwa, C. (1990). Rural-urban linkages in contemporary Harare: Why migrants need their land. *Journal of Southern African Studies*, 16(4), pp. 676–698.

Tawodzera, G. (2013). Rural-urban transfers and household food security in Harare's crisis context. *Journal of Food & Nutrition Disorders*, 2(5), pp. 1–10.

Tawodzera, G., Zanamwe, L. and Crush, J. (2012). The state of food insecurity in Harare, Zimbabwe. *Urban Food Security Series*, No. 13. Cape Town: AFSUN.

Tibaijuka, A. (2005). *The UNHabitat special envoy report on Operation Murambatsvina in Zimbabwe*. Nairobi: UNHabitat.

Toriro, P. (2009). The impact of the economic meltdown on urban agriculture in Harare. *UA Magazine*. Leusden: RUAF.

Zinyama, L. (1993). The evolution of the spatial structure of greater Harare: 1890 to 1990. In: L.M. Zinyama, D.S. Tevera, and S.D. Cumming, eds., *Harare: The growth and problems of the city*. Harare: University of Zimbabwe Publications, pp. 7–32.

Part III

Understanding the urban food systems

12 Food value chains in Kisumu, Kitwe, and Epworth

Environmental and social hotspots

Lesley Sibanda and Harro von Blottnitz

Introduction

The sustainability of food production systems has been a subject of considerable academic and policy interest since at least the time of Malthus. This chapter defines a food system as an "interconnected web of activities, resources, and people that extends across all domains involved in providing human nourishment and sustaining health, including production, processing, packaging, distribution, marketing, consumption, and disposal of food" (Chase and Grubinger 2014: 1). Food systems are made up of individual food value chains which interact with each other through shared infrastructure. They are influenced by various environmental, economic, cultural, social, health, and political factors and therefore also encompass issues of food availability, access, and utilization (Eriksen 2008).

There has been significant focus on the impact of the food system production component on land, water, soil, greenhouse gas emissions, and territorial biodiversity (Westhoek *et al.* 2016). Given the framing of Sustainable Development Goal (SDG) 2 as "End hunger, achieve food security and improved nutrition, and promote *sustainable agriculture*," not "*sustainable food systems*," it is likely that the production component of the food system will remain a key area of research and policy focus. However, there are a number of post-harvest practices that impact food system sustainability and food security, such as food transportation, market structure, and food safety.

In the context of rapid urbanisation, there is increasing focus on urban food security and the impact of urbanisation on food systems. Much of this work has extensively focused on food production in this increasingly globalized agricultural and logistics system; rural and urban linkages; and food access within cities (Tacoli *et al.* 2013). Often, this is either from the perspective of rural populations and the smallholder farmer or the contribution of urban agriculture. More recently, the ways in which this processed food is accessed by the increasing African urban population has become a matter of interest (Crush and Frayne 2011). However, despite this, little research has been done on entire food systems and associated impacts beyond food production and accessibility.

In order to understand the environmental impacts and economic and social rationales of the food systems feeding Kisumu, Kitwe, and Epworth, this chapter employs a hybrid methodology which consists of an adapted environmental and social Life Cycle Assessment (LCA) and hotspots analysis. Life cycle thinking provides a useful and holistic framework for understanding food systems and for evaluating environmental and social effects behind a product. LCA is a technique aimed at addressing the environmental and social aspects of a product and the associated impacts throughout the product's entire lifecycle (Benit and Mazjin 2009). LCAs have been used extensively to assess agricultural systems, specific food processing and manufacturing activities, individual food products, and even whole meals. However, few researchers have attempted to study an entire food system. This is largely due to the complex nature and intrinsic variability of food production systems (Notarnicola *et al.* 2017). Hotspots analysis is a qualitative analysis designed to collate, identify, analyze, and estimate the environmental and social impacts throughout a product's life cycle (Liedtke *et al.* 2010). It is a highly adaptable and flexible methodology which can be used at different scales (i.e. individual products, product portfolios, industry sectors, countries). As such, the hybrid approach aims to identify activities within food systems with significant negative environmental and social impacts (also referred to as hotspots) in Kisumu, Kitwe, and Epworth. This chapter aims to provide an understanding of how these food systems function, and what their sustainability impacts are.

Methodology

To establish how food systems work, an in-depth literature review of academic articles, government publications, and research outputs from various institutions in the field of food production and consumption was undertaken. Next, several signature food value chains were identified (see Chapters 13–15) for investigation and assessment in consultation with local project partners, consisting of staple and commonly eaten foods in each place. The food items selected for investigation are shown in Table 12.1.

This was followed by fieldwork in each case study city. Before commencement of the fieldwork, information sessions were held with research partners and key stakeholders from government, academia, non-governmental organizations, and civil society in each area. This was to determine the relevancy of the

Table 12.1 List of selected foodstuffs

	Epworth (Zimbabwe)	*Kitwe (Zambia)*	*Kisumu (Kenya)*
Staple	Maize	Maize	Maize
Protein	Offal	Fish	Fish
Vegetable	Indigenous vegetables	Indigenous vegetables	Indigenous vegetables
Other	Maputi (popular snack made from maize)	Beans	Eggs
Other	Tomato	Tomato	Tomato

preliminary food items to be tracked (see Table 12.1) as well as to understand which markets to visit.

The fieldwork was then carried out at different landing sites, markets, retail spots, and residential areas. The study was informed by direct observation of various food-related activities and informal interviews and discussions with traders, market leaders, municipal officers, caterers, and residents in the markets and on the streets. Interviews were conducted in a mix of English and the local languages with a local researcher acting as an interpreter. The sites were photographed to document the context for the food system activities. Following the fieldwork, a life cycle approach was used to piece together systems descriptions, paying attention to general features and emerging trends observed repeatedly for the different foods in the different places. Environmental and social impacts in the food systems studied were identified next, using a qualitative Hot Spot Analysis (HSA).

Features and trends observed in the food systems

Based on the literature review, fieldwork, and the analysis of the 15 individual food value chains listed in Table 12.1, a broad picture of the food systems was obtained. Across all study areas, the food systems (comprising multiple interrelated food value chains) were found to be complex and lengthy, and to consist of diverse actors who influence how food is accessed, utilized, and consumed. Food production (agricultural, livestock, and aquaculture and fisheries) is done at both commercial and small-scale levels, the former mostly outside of the studied cities, the latter often within. Some of the locally produced food is sold directly from the producers to consumers, but the majority is sold to processors for processing of food commodities before getting distributed through different channels. Distribution is influenced to an extent by the available transportation infrastructure, trade regulations, and storage requirements. In addition to this, the networks of distribution and retail activities have expanded as markets have globalized. Wholesaling and retail is done through urban markets, supermarkets, kiosks, and street vendors. Consumption happens at different platforms as consumers have access to a diverse range of foods through various retail and catering outlets such as hotels, restaurants, canteens, and street food. The final stage of the food system is disposal, although waste is generated at all levels of the food system. Several key features and trends were repeatedly observed across the different value chains in all the case study areas. These are discussed below.

Imports and exports of food to towns

Increased wealth, urbanisation, and globalization have resulted in substantial changes in the food systems and consumption patterns, with food commodities now travelling to cities, nations, and continents (Westhoek *et al.* 2016). This is largely due to improved transportation, consumer demand, and the implementation of trade agreements that facilitate the flow of food across borders. In each of the case study cities, strong links to local, regional, and global supply chains were evident in both the formal and informal sectors, as across all the cities food from other cities, neighbouring countries, and across the global were traded in both formal and informal markets. As such, consumers were able to access a more diverse range of foods through the multiplicity of retail and catering outlets. Imported foods were at times competing with more locally produced products, while at other times they acted as a complement to domestic production, ensuring that consumers can access certain foods all year that could have been out of season. Often, though, they were providing similar products to different market segments with different demands. For example, in Kitwe the tomatoes sold in Chisokone Market were local and ripened ready for immediate consumption, whereas the tomatoes sold at the supermarket were imported and less ripe, intended for households with storage facilities and less frequent purchasing patterns.

In all the cities, significant quantities of imported foodstuffs from all over the world such as rice, fish, and beans were available in both formal and informal markets. In Kitwe, imported rice from Tanzania was sold by the informal traders while the supermarkets had a wider variety of imported rice from Thailand, Vietnam, Pakistan, India, and from South Africa (which processes imported rice). Similarly, fish from Namibia (mackerel) and China (tilapia) were available for sale in all case study areas. The diversity, richness, and multiple source pathways provide supply resilience, thereby improving food security. However, it is important to also note that some of the food obtained through cross-border trading is undocumented and as such, there are no guarantees on the quality of products.

Market dynamics and affordability

It was also evident that the food systems are in transition with an increased presence of supermarkets in Kisumu and Kitwe. The increase in the number of supermarkets has resulted in changing consumption patterns and has increased the variety of foodstuffs available especially in terms of processed foodstuffs (Stuckler and Nestle 2012). In Kisumu, the informal sector dominates the sale of fresh food, cereals, and indigenous foods, while supermarkets tend to stock and sell more highly processed food. In Kitwe supermarkets, like Shoprite, are now stocking indigenous foods that were previously only sold in the open markets. There are no major supermarket retailers within Epworth. Rather, retail outlets consist of small grocer shops and tuck shops with informal trading being the dominant option for consumers.

An interesting interplay was observed to exist between the formal and informal sectors in Kisumu and Kitwe with consumers buying different commodities from both formal and informal retailers. This was more evident in Kitwe where a Shoprite supermarket was located in the city centre about two blocks from Chisokone Market. At the time of fieldwork for this chapter, informal traders had set up stalls on the pavement outside the front of the supermarket selling indigenous vegetables, fruits, nuts, bread, and various other foodstuffs. The choice of where to buy was thus found to be greatly influenced by access and convenience, as well as affordability. Across all study areas, the formal supermarkets and convenience stores had set prices. In the open markets, it was observed that the prices for some of the commodities vary depending on the quantity of measure. Traders of fresh and perishable foods were more willing to negotiate prices especially towards the end of the day to make sure all stock is sold. Some traders also offered credit facilities thus increasing accessibility of food for the urban poor.

Prevalence of processing

One prominent element of the food systems across the three cities was the practice of processing foods both within the formal and informal sectors. Some processing practices were old and long established, such as the preparation of grain mix for porridge in Kisumu, but others were recent developments responding to new circumstances. The most notable of these was the sale of pre-cooked beans by traders in Epworth in order to reduce the time and energy cost of food preparation for consumers. Another example of this value addition service is the boiling of milk in Kisumu. It is beneficial to the consumer but also has the added benefit of reducing the risk of disease from microbial pathogens that might be in raw milk. Similarly, preservation of vegetables and fish reduces food safety hazards whilst simultaneously increasing the shelf life of the food and minimizing food wastage. Other processing methods that are common are shown in Table 12.2.

Table 12.2 Food processing methods and reasons for processing

Method	*Commodity*	*Reason for processing*
Milling	Maize	Convenient for the consumer.
	Grain mix for porridge (millet, sorghum, and cassava)	Convenient for the consumer and reduced meal preparation.
	Okra Tomatoes	Mainly done so that the commodities are available all year to the consumer.
	Peanuts (peanut powder)	Convenient for the consumer and reduced meal preparation.
Drying	Fish Indigenous vegetables Mushrooms Caterpillars	To extend the shelf life through preservation, to minimize food losses, and to ensure foodstuffs are available all year round.
Smoking	Fish	
Frying	Fish	
Pre-boiling	Beans Milk	Reduced preparation time for the consumer and less energy consumption during preparation.
Cooking	Porridge Meals Snacks (scones, samosas, and salted boiled peanuts)	These street foods offer consumers affordable and quick meals especially for people with busy lifestyles.

Hotspots along the food value chains

Food systems have become increasingly complex as seen from the multiple ways of accessing food. Access to nutritious food is central to alleviating poverty and improving living standards, but food production, processing, trading, and consumption can also have negative social and environmental impacts. Identifying these negative impacts in food systems can support decision making on strategies for improving their resilience as well as highlighting the entry points for appropriate interventions. Thus, the study conducted an environmental and social analysis across the individual value chains in an attempt to identify the social and environmental issues that could impact food systems. The results show that hotspots occur in different phases along the value chain. In the environmental analysis the major "out-of-town" environmental hotspot identified was deforestation linked to biomass cooking fuels. Waste disposal is also seen as a hotspot due to the poor waste management practices across the value chains. The social analysis revealed workplace safety and work conditions as potential threats. Other identified hotspots include consumer health, food safety, and product safety. Three case studies (one from each case study city) are used to illustrate the possible hotspots that occur along each value chain in Figure 12.1.

While the study identified several challenges across the entire value chains as shown in the figure above, this section will discuss three of the impacts. The selected ones occur across all cities in the value chains examined.

Inadequate waste disposal and poor resource efficiency

Disposal of waste generated at all levels of the food system indicates that there is an interface here with waste management systems. Observations indicated that solid waste management is a pressing issue and a significant environmental and public health challenge in all the case study cities. In Kisumu and Kitwe, only 20–25% of the waste generated is adequately collected and properly disposed of at landfills (Gutberlet *et al.* 2017; Magezi 2015). These waste collection services are provided by both the government and the private sector and mostly concentrated in middle- and high-income areas (Gutberlet *et al.* 2017; Magezi 2015). In Epworth, solid waste management is given low priority with very limited collection services by the Epworth local board.

Across all study areas, some organic waste, which comprises about 60% of the total waste generated, is collected and used as animal feed by pig and rabbit breeders (Magezi 2015). It is also used to make compost which is then used in home gardens and in other instances sold on to other users. The rest of the uncollected organic waste is disposed of by throwing it on the ground or in rubbish pits thereby posing a health risk especially in the rainy season as it creates a favourable environment for the growth of bacteria and microbial pathogens.

In Epworth and across low-income areas in Kisumu and Kitwe, the inadequate waste collection services have triggered backyard burning and widespread illegal open dumping along passageways, roads, and empty plots of land. The open burning of waste results in air pollution. The toxic pollutants emitted can adversely affect human health, thereby making the urban poor more vulnerable. Across all three study areas a significant environmental impact arises

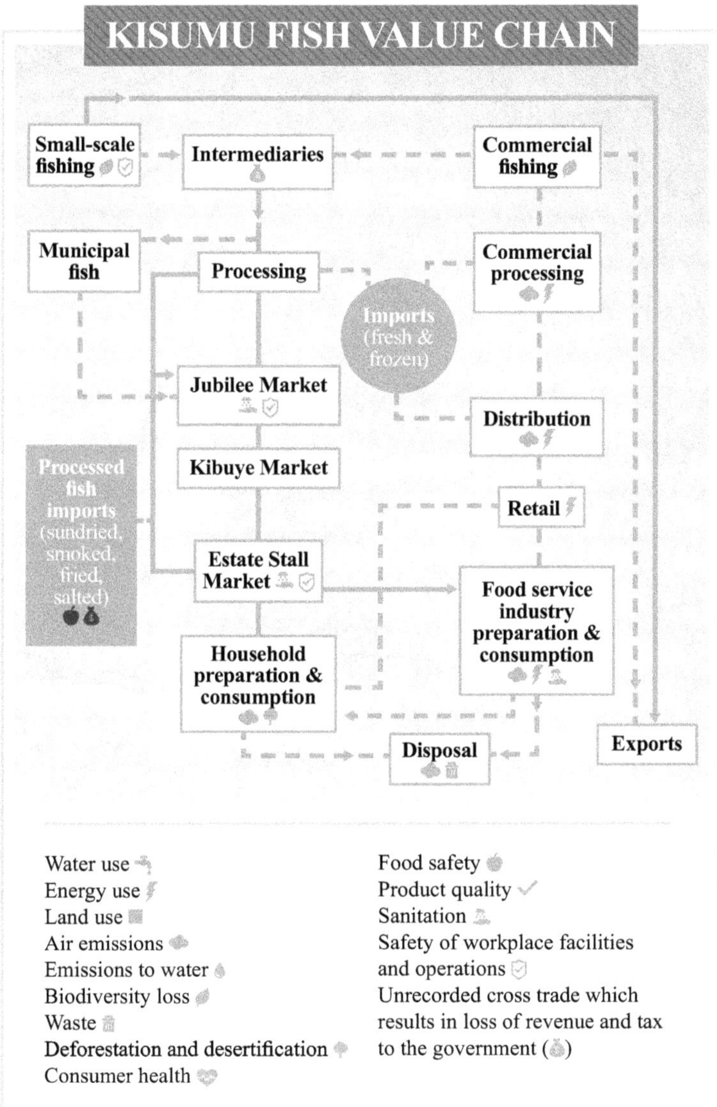

Figure 12.1 Illustrations of the food chain with hotspots shown

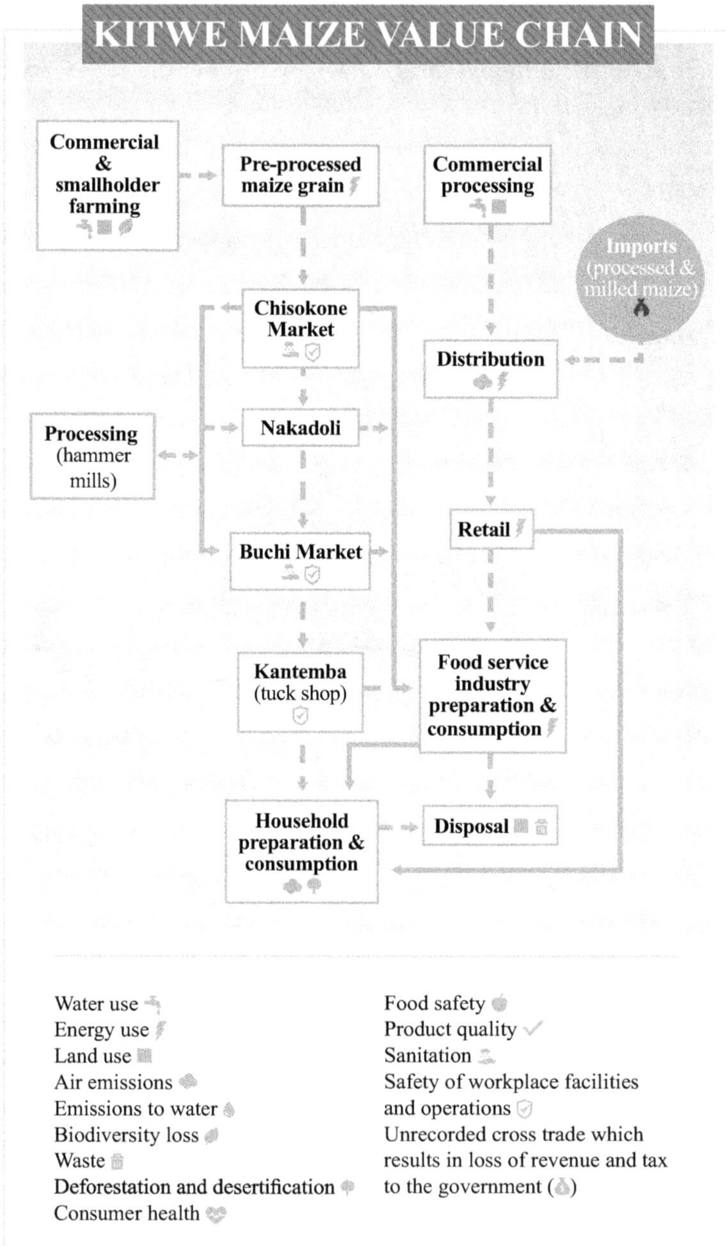

Figure 12.1 Continued

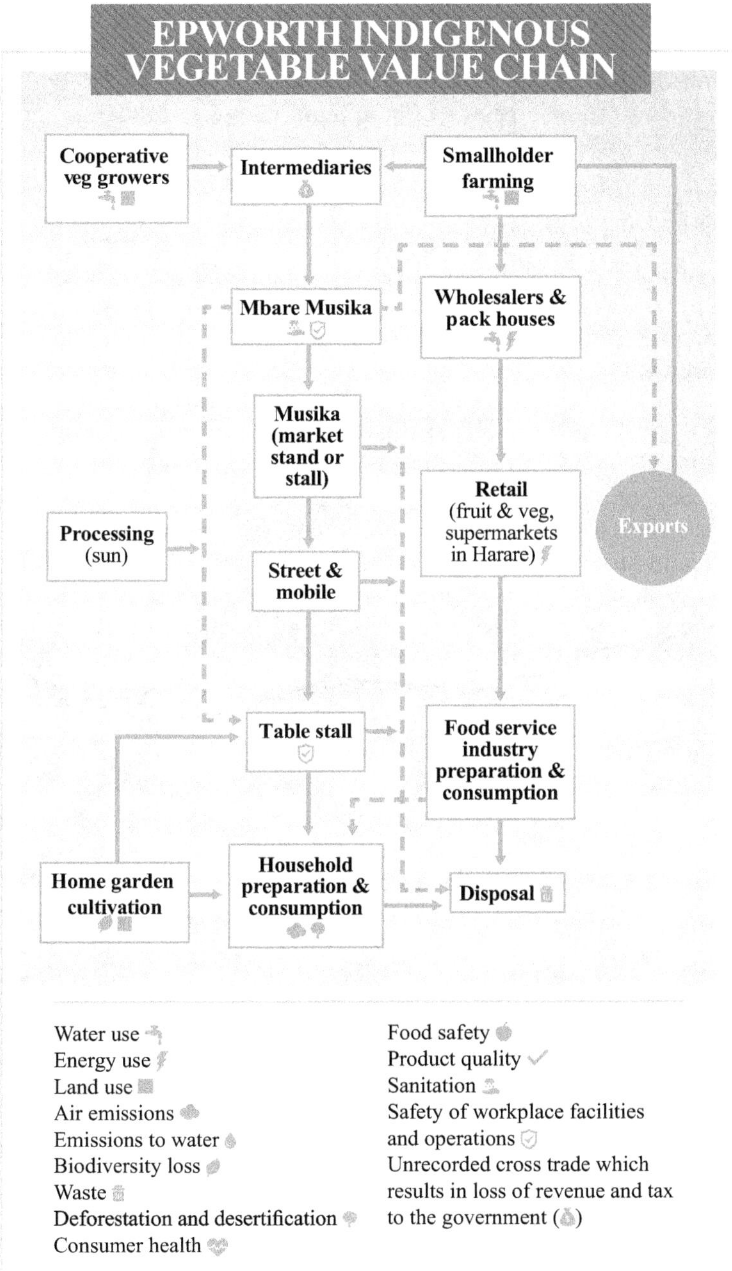

Figure 12.1 Continued

from the illegal dumping as well as from the non-sanitary conditions at the few available landfill disposal sites.

Risks of open dumping include soil pollution, underground water contamination from the leachates from the dumping sites, bad smells, and fire outbreaks. In addition, these dumping piles are breeding grounds for flies, rats, rodents, and mosquitoes and can be a source of communicable diseases such as fever, diarrhoea, dysentery, cholera, and malaria (Okot-Okumu 2012).

Environmental and social implications of biomass fuels

Biomass fuels are the predominant energy source used for cooking and preservation by smoking in all the three case study areas. In Zambia, biomass fuels consist mainly of firewood and charcoal, accounting for approximately 70% of the country's energy needs (Giesecke 2012). Similarly in Kenya, firewood and charcoal are the commonly used biomass fuels and cater to approximately 80% of the Kenyan population's energy needs (Institute of Economic Affairs 2015). The energy mix for Epworth is composed of mainly sawdust and firewood with charcoal being used by the minority. In both Kisumu and Kitwe, charcoal is the preferred choice over firewood as it is easier to transport and store, burns for longer, and can be easily extinguished and reheated again. In addition, the relatively low cost of biomass fuels compared to electricity, liquid petroleum gas (LPG), and kerosene make it a preferred fuel source especially for the urban poor. The high demand of biomass has led to the unsustainable exploitation of forests and woodlands resulting in soil erosion, forest degradation, and deforestation (Giesecke 2012; Institute of Economic Affairs 2015). In Zambia alone, charcoal production contributes to deforestation by an average loss of over 300,000 ha of woodland annually (Giesecke 2012). Some of the impacts of deforestation include the loss of habitats from the disturbance of forests, changes in species diversity, depletion of water sources, and the destruction of water catchments (Institute of Economic Affairs 2015). Initiatives to minimize the impacts have been implemented in both Kenya and Zambia, but the firewood and charcoal production sectors are entirely informal and thus the enforcement of regulations poses a serious challenge.

In the food sector, biomass fuels are also used to preserve different food products through drying and smoking as well as general food preparation. In a lot of cases, biomass fuels are burned in open fires or stoves with poor combustion characteristics. As a result, the wood fuel is not completely and properly combusted leading to indoor air pollution from harmful emissions and pollutants such as carbon monoxide, formaldehyde, and various other hydrocarbons and compounds (Smith 2006). The prolonged exposure to these pollutants and high concentrations of smoke constitutes a significant public health hazard and can lead to chronic respiratory disorders such as asthma, emphysema, chronic bronchitis, tuberculosis, chronic obstructive pulmonary disease (COPD), and in more extreme cases lung cancer (Smith 2006). The extent of these disorders is largely determined by both the pollution levels and the time spent breathing the polluted air.

To mitigate these heath impacts, the complete combustion of biomass fuels is imperative as well as the introduction and implementation of appropriate safety measures and interventions designed to reduce the long-term impacts of biomass smoke exposure. This can be achieved through efficient ways of burning the biomass fuels and proper ventilation of the cooking facilities used during the smoking business.

Improving the efficiency of the stoves is another possible intervention, but this needs to take into account the cultural, social, and economic significance of cooking using wood fuels. The use of wood fuel for cooking is determined by cooking traditions that have been passed down from previous generations as well as the flavour that is imparted to food from the choice of biomass fuel used. Other suggestions to mitigate the impacts is to shift from using biomass fuels to other low-emission fuels such as LPG and kerosene, especially at the processing stage.

Health, safety, and security hazards in the markets

Foodstuffs such as meat, fish, and milk are traded in most local markets, which often lack modern infrastructure, and thus can potentially pose significant health risks. Most of the markets visited have inadequate or no electricity. Waste disposal facilities, water supply, and sanitation facilities are also inadequate leading to unhygienic practices. Due to inadequate infrastructure, traders have makeshift trading spaces with locally designed canopies, and in some markets the stalls are open and uncovered and traders simply spread their wares on the ground and sell under umbrellas. This puts the traders at risk as they are exposed to heat stresses especially during the dry and hot season. Similarly, in the rainy season, trading can be disrupted leading to product loss and/or infestation of their commodities.

To compound this, most markets have no refrigerators, insulated containers, or ice available for the traders to store fresh produce especially fish and meat. As a result, the produce is often left exposed to weather elements for long periods of time with flies constantly hovering over it. Exposure to direct sun rays and increasing temperatures throughout the day increase the risk of food deterioration and the food is then sold at cheaper prices at the end of the day resulting in loss of income for the traders. In certain instances, some traders refreeze the leftover produce at home and resell it the next day creating a significant health risk. To minimize the associated health risks, sufficient toilet facilities need to be provided in the markets with clean and reliable water supply for flushing, handwashing, and personal hygiene. The provision of insulated containers, ice boxes, and covered food stalls can minimize hazards by ensuring the food is kept fresh and not exposed to flies and other insects.

Security is another major challenge as traders have to constantly be alert to theft of their goods especially overnight. Lastly, fire outbreaks are another risk and concern for the traders. Fire outbreaks are a common occurrence in the markets, resulting in traders losing all their merchandise. The major causes of fire include illegal electricity connections and the choice of materials used for construction of the market stall structures, which are often highly combustible. This, coupled with the fact that there are no visible hydrant points and fire extinguishers in markets, makes it easy for fire to spread throughout the market with devastating consequences. The provision of fire extinguishers to traders as well as proper electrical connections would help mitigate these outbreaks thereby minimizing loss for the traders.

Conclusion

The descriptions and analyses presented in this chapter indicate that urban food systems are complex and changing. In Kisumu and Kitwe, the systems are more diverse with a wider variety of foodstuff. Epworth differs slightly, with less variety of food produce being marketed. This can be attributed to its being a satellite settlement while the others are significant cities.

Across all three cities, processing of foodstuffs is prevalent. It is undertaken to minimize food wastage and reduce mealtime preparation and energy consumption for the consumers. Many traders consider this as a value-added service, especially for consumers who have time constraints. Food is sold through a wide variety of formal and informal food retail outlets including supermarkets, open traditional markets, grocers, kiosks, roadside stalls, and street food vendors. Both the formal and informal food sectors play an important role in ensuring the urban poor can access food. The food value chains extend beyond the cities and have district, national, regional, and continental linkages. Food imports and exports are a prominent feature of the food systems with a diversity of imported food commodities available in formal and informal markets.

The analysis revealed several environmental and socio-economic hotspots. The use of biomass fuels in food processing is associated with challenges in biomass fuels production and use, many of which have been documented. These include deforestation, desertification, and air pollution and associated respiratory disorders. Linkages between food processing and environmental impacts have also been observed. Thus, despite offering a way of reducing food waste, common food processing methods have been shown to sometimes lead to profound environmental and health impacts.

Waste disposal is another identified hotspot, with insufficient solid waste management services in all the cities. This leads to environmental degradation and poses a significant health risk to residents. Lastly, food and general safety is of key concern, as most markets do not have sufficient water or energy supply or sanitary facilities. The highlighted environmental and social impacts may not always be a direct threat to the survivability of food systems (and thus to food security), but certainly impact negatively on development outcomes. An understanding of these impacts is therefore important as it highlights opportunities for improving resource efficiency within food systems and identifies stages within the food systems that could benefit from effective policy interventions.

References

Chase, L. and Grubinger, V. (2014). *Introduction to food systems: Food, farms and community: Exploring food systems*. Lebanon, NH: University Press of New England.

Crush, J.S. and Frayne, G.B. (2011). Urban food insecurity and the new international food security agenda. *Development Southern Africa*, 28(19), pp. 527–544.

Eriksen, P.J. (2008). Conceptualizing food systems for global environmental change research. *Global Environmental Change*, 18(1), pp. 234–245.

Giesecke, C. (2012). *Deforestation in Zambia: Driving factors and potential solutions*. Washington, DC: United States Agency for International Development.

Gutberlet, J., Kain, J.H., Nyakinya, B., Oloko, M., Zapata, P. and Campos, M.J.Z. (2017). Bridging weak links of solid waste management in informal settlements. *Journal of Environment & Development Policy Review*, 26(1), pp. 106–131.

Institute of Economic Affairs, (2015). *Situational analysis of energy industry, policy and strategy for kenya*. Nairobi: Institute of Economic Affairs.

Liedtke, C., Baedeker, C., Koleberg, S. and Lettenmeier, M. (2010). Resource intensity in global food chains: The hot spot analysis. *British Food Journal*, 112(10), pp. 1138–1159.

Magezi, T.J. (2015). *An investigation of the imbalance of a fast-growing consumer culture and insufficient waste management infrastructure across a number of sub-Saharan African Cities*. Masters, University of Cape Town.

Notarnicola, B., Sala, S., Anton, A., Mclaren, S.J., Saouter, E. and Sonesson, U. (2017). The role of life cycle assessment in supporting sustainable agri-food systems: A review of the challenges. *Journal of Cleaner Production*, 140(2), pp. 399–409.

Okot-Okumu, J. (2012). Solid waste management in African cities—East Africa. In: L. Fernando Marmolejo Rebellon, ed., *Waste management—An integrated vision*. Rijeka, Croatia: InTech, pp. 3–21.

Smith, K.R. (2006). Health impacts of household fuelwood use in developing countries. *Unasylva—An international journal of forestry and forest industries*, 57(2), pp. 41–44.

Stuckler, D. and Nestle, M. (2012). Big food, food systems, and global health. *PLoS Medicine*, 9(6), p. e1001242.

Tacoli, C., Bukhari, B. and Fisher, S. (2013). Urban poverty, food security and climate change. *Human Settlements Working Paper 37: Rural-Urban Interactions and Livelihood strategies*. London: International Institute for Environment and Development (IIED).

Benit, C. and Mazjin, B. (eds) (2009). *Guidelines for social life cycle assessment of products*. Paris, France: UNEP/SETAC Life Cycle Initiative, Sustainable Product and Consumption Branch.

Westhoek, H., Ingram, J., Van Berkum, S., Özay, L., and Hajer, M. (eds) (2016). *Food Systems and Natural Resources. A Report of the Working Group on Food Systems of the International Resource Panel*. Paris, France: United Nations Environment Programme.

13 The characteristics of the urban food system in Kisumu, Kenya

Paul Otieno Opiyo and Harun Okello Ogindo

Introduction

This assessment of Kisumu's food system seeks to scrutinize food poverty as an indicator of the workings of persistent, pervasive, and systemic urban poverty. The urban food system includes the whole pathway and nodes through which food flows in the value chain. It includes an understanding of food production sources, food processing, distribution, marketing, and consumption patterns, as well as other socio-economic and environmental inputs and outcomes.

Background to the historical and economic development of Kisumu as well as its size and growth and spatial change is contained in the Introduction to the book. The city is home to particularly high levels of food poverty (53%) when compared to Nairobi (8.4%) and Mombasa (38.6%) (Mireri *et al.* 2007). This chapter will describe Kisumu's food system, detailing the sources of food, the characteristics of food retail, and the role of the food system in livelihoods and food security. The chapter will further discuss food retailers' responses to customer needs and challenges to food retail in the city.

Research methods

The Consuming Urban Poverty (CUP) project used the following research methods to understand urban food security and the urban food system in Kisumu.

Food retail mapping and survey

An initial mapping of food retail sites, typologies, and trends in Kisumu formed the basis for the sampling process for a more detailed food retail survey. Four sites in the city were selected for mapping. These were: 1) Nyalenda Informal Settlement (chosen because of its position as one of the largest informal settlements in the city, housing residents of mixed income categories); 2) Jubilee Market (the main municipal market, which includes a fish market and many roadside traders around the main bus station); 3) Kibuye Market (the largest market in the city); and 4) Oginga Odinga Street (the main street in the Central

Business District (CBD)). The mapping exercise was done in each of the sites on specific days between 30 May and 10 June 2016. The mapping was designed to give an indication of the following typologies and trends: the gender of store owners or operators, the location of the stores, the nature of a store's operations, and a store's typology and product mix. The data was collected using handheld tablets that also facilitated collection of the geo-coordinates of the 2167 retail outlets mapped in the four sites.

The mapping formed a basis for sampling of key food retail outlets for a more detailed food retail survey. For the population of 2167 food retailers, and allowing for a margin of error of 5% at a 95% confidence level, the sample size required was 327. Field testing of the survey tool confirmed that a 90% response rate was achievable, and therefore at least 363 food retailers were surveyed. Due to the diversity of types of food retail outlets and the targeted sites, it was necessary to adjust the number upwards to 550 retail outlets. In September and October 2016, 551 food retail outlets were surveyed. In each of the sites, the respondents were selected using purposive sampling, ensuring that the whole area was covered and there was diversity in the typologies of food retailers; and respondents were proportionately distributed across the study sites as shown in Table 13.1. The process involved an in-depth survey into a variety of key food system and food retail operating processes and the retail environment.

Five-item food value chain survey

This survey sought to test the widely held assumption that urban areas receive food from agricultural activities in the immediate surroundings. It traced five key food items widely consumed in the city: *ugali* (maize or sorghum or millet meal), fish, vegetables, porridge, and eggs. The items were traced from retail outlets to source rather than in the conventional direction of "farm to fork". The survey adopted purposive sampling, where specific retail points were deliberately selected to ensure variety in terms of the typology of the sales outlets. The food-item-source enquiry was done in October 2016 and involved a deliberate engagement with the identified retail outlets to enquire as to the source of the specific product. We also recorded the contact details of the supplier to that retail outlet. The process continued until either no further information was

Table 13.1 Distribution of food retail survey respondents

Site	*No. of food retailers mapped*	*No. of food retailers surveyed*
Nyalenda Informal Settlement	985	231
Oginga Odinga Street	129	33
Jubilee Market	413	100
Kibuye Market	640	187
Total:	2167	551

available or the original source was identified. Similar questions were asked of the different suppliers as the enquiries retreated along the value chain.

The nature of the urban food system in Kisumu

Sources of food traded in Kisumu

The food system of any city is a hybrid – it combines different means of food provisioning and consumption. Some cities mainly rely on urban, peri-urban, and nearby rural farms and food processors, while others depend mostly on food produced and processed in other countries or continents (FAO 2016). Kisumu is deficient in food production and food processing industries. However, food is available through imports from other counties in Kenya and from other surrounding countries. The CUP reverse value chain analysis of five key food items established that an insignificant proportion of the food consumed in Kisumu is produced in the immediate neighbourhood. The five food items traced were *ugali*, fish, vegetables, porridge, and eggs. The location of the sources of production of these food items in Kenya is illustrated in Chapter 8, Figure 8.1.

Ugali (maize/sorghum meal)

Ugali is the staple food of most residents of Kisumu. Most of the maize meal consumed in the city comes from Trans-Nzoia, Uasin-Gishu, Narok, and Bomet counties in Rift Valley, as well as Busia (Kenya) and Busia (Uganda). These areas are more than 100km away from Kisumu. Sorghum for making brown *ugali* is brought into Kisumu mainly from Busia (Uganda). Middlemen are involved in sourcing the grains from farmers in these areas and delivering it to markets and millers (see Figure 13.1).

Fish

The three most popular types of fish consumed in Kisumu are the silver cyprinid (*Rastrineobola argentea*), the Nile tilapia (*Oreochromis niloticus*), and Nile perch (*Lates niloticus*). Because Kisumu is located on the shores of Lake Victoria, it is widely expected that fish consumed in the city comes from the beaches around the city. However, it was established that fish sold in Kisumu is imported from beaches further away, as fish stocks in the gulf around Kisumu are declining due to pollution and the massive growth of water hyacinth linked to rapid expansion of the city without provision of appropriate infrastructure.

Tilapia and Nile perch are imported into the city from areas more than 70km away. The main sources are Yimbo, Bondo, and Rarieda in Siaya County; Migingo Islands; and Busia (Port Victoria) and Busia (Uganda). Dried fish from Lake Turkana and smoke fish from Busia (Kenya and Uganda) was also found in the markets. Frozen tilapia from China is imported in containers and distributed by middlemen to markets and restaurants in Kisumu. Silver cyprinid is imported from Mbita and Sindo in Homa Bay County, Sori in Migori County

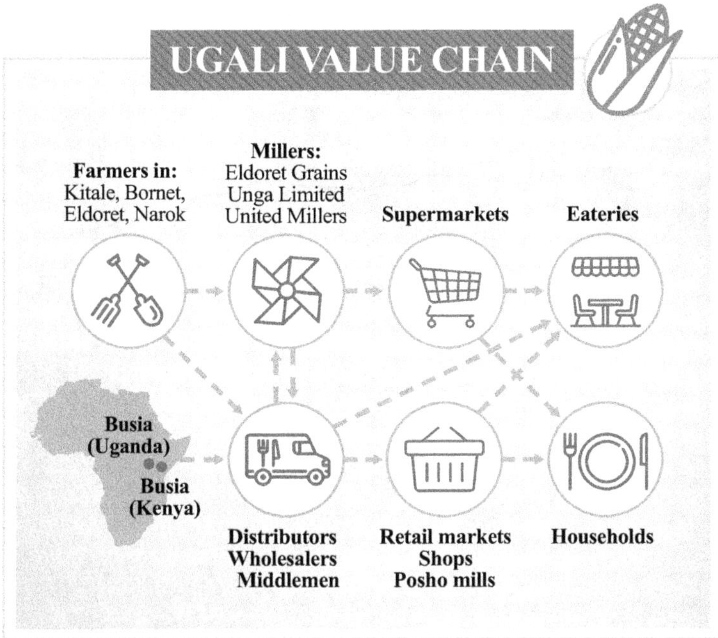

Figure 13.1 Value chain for *ugali* (maize meal)

Note: Eateries includes hotels, restaurants, and food kiosks

(CUP Five-Item Food Value Chain Analysis 2016)

(more than 120km away), Port Victoria in Busia, and Usenge and Luanda in Siaya County. Processed and packaged fish fillets found on the supermarket shelves in Kisumu were traced to two companies based in Nairobi – Alpha Foods and Farmers Choice (see Figure 13.2).

Vegetables

Cabbage and kale are imported from Bomet, Kisii, and Nakuru Counties. They are also seasonally brought into the city from Sondu and Kano within Kisumu County. Traditional vegetables seasonally come from Kano and Kajulu in the peri-urban areas of the city. However, more reliable sources are Vihiga County and Kisii County, which are 30km to 100km away. Traders in Kisumu buy the vegetables in bulk from Jubilee and Kibuye markets and sell in smaller units in other markets within the city and peri-urban areas (see Figure 13.3).

Porridge

Porridge is a popular food item made from a mixture of millet (the main ingredient), sorghum, cassava, and maize flour. Millet is mainly imported from Busia

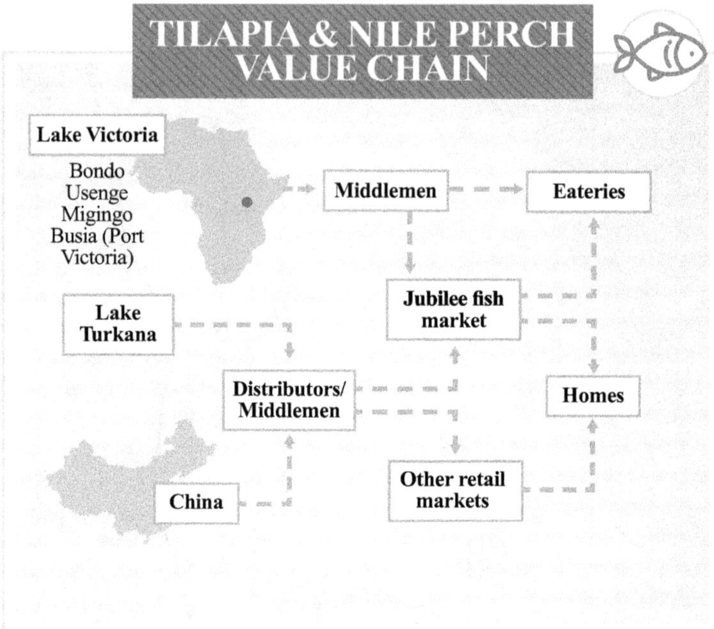

Figure 13.2 Value chain for tilapia and Nile perch

(CUP Five-Item Food Value Chain Analysis 2016)

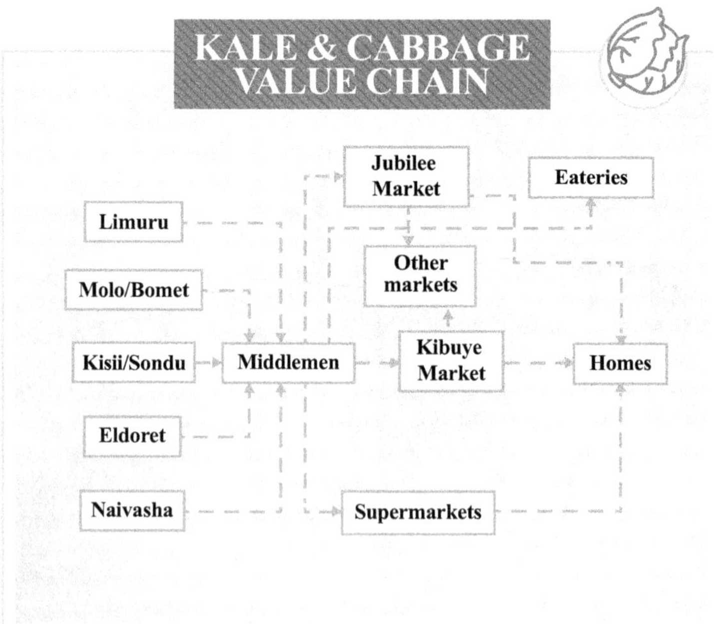

Figure 13.3 Value chain for kale and cabbage

(CUP Five-Item Food Value Chain Analysis 2016)

(Kenya and Uganda), Kisii County, and Migori County (Awendo area). The porridge vendors get supplies from Kibuye Market for milling in small-scale grinding mills, where it is possible to get smaller amounts of supplies milled (see Figure 13.4).

Eggs

The origin of eggs sold in supermarkets and municipal markets was traced to Uganda, Nakuru County (Nakuru and Naivasha towns), and the Kibos area on the outskirts of Kisumu City. The poultry farm in Kibos is the main supplier to supermarkets and eateries in Kisumu but it is unable to meet this overall demand. The company therefore imports eggs from Uganda and also collects eggs from small-scale farmers to supplement their production. Interviews with dealers revealed that eggs from Uganda are cheaper in Kisumu because of readily available and cheaper animal feeds in Uganda (see Figure 13.5).

Characteristics of food retail in Kisumu

Food retail in Kisumu is largely informal. Formality or informality of food retail in the city is defined by the kind of trading premises (whether permanent or temporary); location of the business (in approved market places or unapproved markets on the roadside); and the legal status of the business enterprise (whether it is registered/licensed, or unregistered/unlicensed). Apart from the established municipal markets, food retail often takes place in temporary structures and in unapproved markets on the roadsides. The CUP food retail survey established that although a big proportion (57%) of food retailers are licensed,

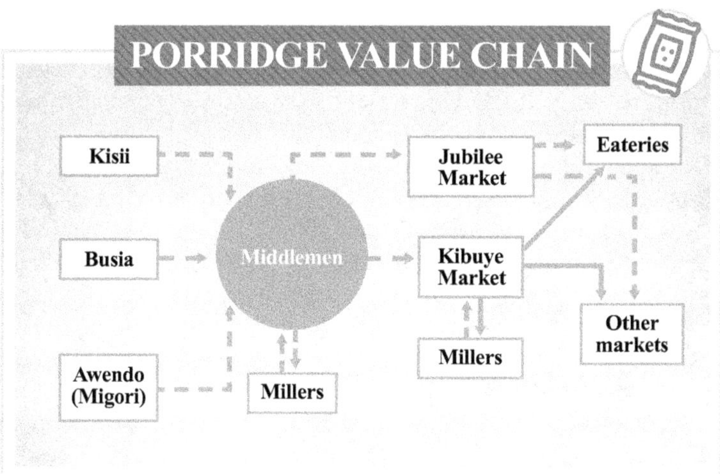

Figure 13.4 Value chain for porridge

(CUP Five-Item Food Value Chain Analysis 2016)

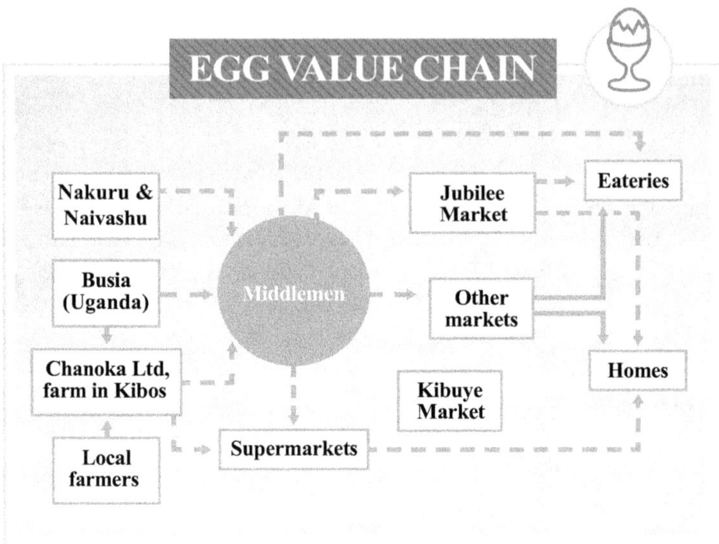

Figure 13.5 Value chain for eggs

(CUP Five-Item Food Value Chain Analysis 2016)

a significant proportion (19%) trade without licenses, a further 17% are unregistered businesses that pay daily vending fees, and 7% indicate that they did not require licenses or they paid market operators. This informality is driven by the high levels of poverty and unemployment. Informal food trade is a source of livelihood for many and also a source of food for the majority of the poor city dwellers because they provide accessible and affordable food, sometimes on loan and in small quantities (IIED 2016; Skinner and Haysom 2016).

While all municipal markets are specially built and designated by government authorities, they do, however, vary in size, supply, and formality. Most people in Kisumu depend on the markets for their daily supply of groceries. The CUP household survey (discussed in detail in Chapter 16) established that 64% of households in surveyed areas obtain food from the markets at least once a week. Two main markets – Jubilee and Kibuye – hold great significance for local food retail as they are the main source of supplies for the smaller markets and informal traders in the city as well as suburban markets.

Other than in the markets, informal food retailers are found on the streets, in informal settlements, and in the CBD. The CUP retail mapping established that 28% of retailers in the selected areas traded on the street edge, and a further 8% were mobile, moving from street to street with their merchandise in disregard of city authority regulations. Street-edge food retail is most common in informal settlements and at the main transport confluences connecting residential areas in the city. These include the main bus terminus in the city centre and public transport (*matatu*) termini in residential areas. In many instances, these temporary food stalls (*vibanda*) are demolished to facilitate road construction or repair works.

The locations of food retail businesses are determined by a number of factors which were uncovered in the CUP food retail survey. It was found that more than 27% of food retailers in permanent outlets traded at their locations because space was available; 26% traded at their locations because they seized a perceived business opportunity; 18% chose their trading places due to affordability; 16% chose their trading places due to other economic factors; and 13% traded at their premises due to seasonality, a specific event or permit. Because of these factors, food outlets are concentrated in the major markets (where there are permanent structures) and in residential neighbourhoods and on the streets (targeting residents and passers-by along the streets).

The food retailers in Kisumu were found to sell a wide variety of foods. The most traded food products were fruits (26% of cases), vegetables (25%), sugared drinks (20%), and eggs (19%). These are closely followed by dry grains (16%), snacks (chips and crisps) (15%), bread (15%), fried foods (13%), and sweets (13%). The least traded products were smoked and salted fish and frozen fish (less than 1%), live chickens (1%), non-cooked meals (2%), frozen meat (2%), and fresh fish (4%). These findings suggest that the most traded products are those that are ready to eat or cheaper and easier to prepare, and ready snacks. The least traded are those foods that require more energy to cook and/or preserve. As discussed in Chapter 16, food availability and access for the poor is limited by access to energy and storage facilities.

The last two decades have witnessed the growth of supermarkets in Kisumu's CBD and more recently into residential neighbourhoods. Interviews with the local chapter of the Kenya National Chamber of Commerce and Industry revealed that in the early 1990s, there were only two supermarket brands in Kisumu – Yatin and Foamat – and by 2016, there were eight major Kenyan supermarket brands, with a total of 15 branches in Kisumu. The formal retail sector is clearly growing rapidly, but evidence from the CUP survey indicates that supermarkets have not yet become a significant contributor to household food provisioning. The CUP survey found that 55% of sampled households purchased less than 25% of their food from supermarkets. The CUP household survey revealed that about 40% of households in the suburbs surveyed buy some of their food from supermarkets once a week, 48% do so once a month, while 69% of households obtain food from street vendors at least five days in a week and 82% of households obtain food from kiosks and home shops at least five days a week.

The role of the food system in livelihoods and food security

Unemployment in Kisumu is high at 30%, which is significantly higher than the national average of 13% (Mireri *et al.* 2007). The informal sector employs about 52% of the city's working population (and up to 70% of the working population in informal settlements) and supplies on average a monthly income of about KES 4000 (US$40) (UN-Habitat 2005). As Table 13.2 indicates, employment is very low in the food retail sector, as it shows that only 6% of food retail businesses have between one to ten permanent employees, and only 1% has more than 11 permanent employees. Emphasizing the lack of formal employment is the fact that only about 15% of food retail businesses employ between one and five casual workers, and that 93% of food retailers have no permanent staff employed.

The CUP food retail survey also revealed that the food retail sector is primarily a source of livelihood for individual small-scale owners, as 90% of them are owner-operated and only 10% are operated by employees. Ten percent of those who are employed in the sector are casual workers, earning an average of KES 4000 (US$40) per month. The CUP retail survey also established that the food retail sector is dominated by women – about 70% of food retail operators were female. Women not only dominate the food retail sector, but older women tend to dominate involvement compared to older men. This was illustrated by the results that show very few males participated in the retail business between the ages of 51 and 60, and none at all above 60 years, while in contrast, female participation continued up to 60 years and older. This confirms the contention that women engage in food retail as a permanent livelihood activity as opposed to temporal, survivalist engagement (Skinner and Haysom 2016).

For the urban poor, street vending is one of the primary means of earning a livelihood, as it requires minor financial input and the skills involved are low. A particularly large section of street vendors in urban areas are low-skilled migrants from rural areas in search of employment (Onyango *et al.* 2012). Traders in the food retail sector support many dependants through their business. The CUP food retail survey confirms that traders in the food retail sector support many dependants through their business as it shows that 63% of traders support one to five household members, 35% support six to ten household members, and 2% support more than ten household members. Due to the poor state of the economy and high rates of unemployment in Kisumu, food retail becomes one of the first economic opportunities perceived by those who strive to earn a living, as expenditure on food is a priority to all, especially poor households.

Table 13.2 Employment in food retail business

Permanent	*%*	*Casual*	*%*
Between 1 and 5	5	Between 1 and 5	15
Between 6 and 10	1	Between 6 and 10	1
More than 11	1	More than 11	1
No permanent staff employed	93	No casual staff employed	83

(CUP Food Retail Survey 2016)

Food retailers' responses to customer needs

Customer needs in the food retail sector are complex and dynamic, primarily being dependent on the social classes within the city. Food retailers in Kisumu have adopted different strategies to cope with needs of customers, including selling ready-to-eat packaged snacks, bulk breaking, offering credit facilities, selling cooked foods, and roadside catering.

Small-scale retailers buy food in bulk from the main markets and sell in smaller units in informal markets. For example, even though fruits and vegetables were found to be cheaper per unit in supermarkets, poor residents of informal neighbourhoods generally do not buy these from supermarkets but from informal traders on the roadsides who sell in smaller units. It is common to find informal traders selling sugar, milk, cooking oil, and maize meal in smaller measures than would ordinarily be found on supermarket shelves. These small unit measures are not meant to make items cheaper but rather to enable households with low daily incomes to purchase food – this is locally referred to as *kadogo economy*. Food retailers also offer credit to attract and retain customers. The CUP food retail survey found that 35% of them regularly offer credit facilities and 13% occasionally offer it. Of those who do offer credit, 95% of cases indicated that they offer credit to retain customers. Because one of the ways through which poor households navigate the food system is by borrowing, credit sales can therefore be seen as a direct response to the needs of clients.

Food retailers are also located in areas that are easily accessible to clients. It is for this reason that three of the retail markets built in Kisumu under the slum upgrading program – Migosi Market, Manyatta Peace Market, and Kowino Market – remain underutilized, as traders tend to move to the roadsides for easy access by customers. According to the city management over 10,000 hawkers operated in undesignated areas in 2015 and hence they regard the city as "disorganized". The existence of hawkers on the streets is a direct response to the needs of their clients, who purchase merchandise conveniently along the roads or next to public transport termini. Legitimate (licensed) traders, who pay land rates, electricity bills, workers, and other county charges, have raised concerns over insecurity, unfair competition, and unfair treatment by the hawkers who sell similar goods right at their door steps. Attempts to relocate traders into designated trading areas so as to avoid these conflicts include the planned reconstruction of three markets that are currently in use at Jubilee, Otonglo, and Kibuye (Omollo 2015).

Challenges to food retail in Kisumu

Kisumu's rapidly expanded population since the late 1960s and its high poverty levels have resulted in demands for cheaper and easily accessible food. New businesses are proliferating in response to the growing demand for food by the increasing population, and with this rapid growth, challenges in the sector have escalated. In this chapter these challenges will be classified as infrastructural, governmental, financial, and economic.

Lack of or inadequate infrastructure is the most important challenge to food retail in Kisumu. Access to infrastructure is determined by governance at the national and county levels. Kisumu used to be an important rail and water transport node but now its main transport links are by road. Interviews with city officials revealed that only an estimated 30% of the existing road networks in the city were being maintained in 2016. Broadly speaking, poor road networks increases the cost of transportation of food, leading to higher food prices in the city, and therefore limited access to food, particularly by the poor. The CUP food retail survey confirms this idea, singling out transport as the highest cost to retailers other than stock (82% of cases) in food retail. This is due to the poor road network in Kenya and the distance from the food sources. Spoilage (51% of cases) is the second highest cost after transport, which the survey found to be largely attributed to the lack of basic infrastructure to prevent food spoilage: 54% of cases lacked electricity, 75% lacked refrigeration facilities, and 94% lacked air conditioning facilities. Other important infrastructural facilities were equally lacking or inadequate. For example, 92% lacked designated parking facilities; 61% lacked secure storage; 70% lacked piped running water; and 57% lacked sanitation facilities. These infrastructural inadequacies reduce accessibility to food and raise food safety concerns.

As noted in Chapter 8, governance of food systems is a key challenge to food retail in Kisumu. The licensing procedures are simple for formal food retailers, but due to the difficulty in complying with food hygiene standards and financial costs, informal traders tend to avoid licensing and therefore inspection. The CUP food retail survey established that 34% of food outlets are never inspected, either because traders avoid inspection or because of inadequate capacity of the city authorities. Food retailers adopt different tactics to avoid inspection; for instance, 13% avoid inspections by paying bribes and another 10% avoid inspections by closing until inspectors move on. Despite often managing to evade inspections, street vendors and roadside traders are often evicted from their trading places, resulting in violent confrontations with local government agents and the police. Despite being good for the environment, the recent government-imposed ban on the use of plastic bags for packaging poses an additional challenge to both food retailers and consumers as the available alternatives (both in terms of alternative packaging materials for the same foods and alternative pre-packaged foods) are more expensive, meaning consumers have ended up paying more for food. This ban has also unintentionally resulted in some informal traders using unhygienic packaging materials. This suggests that government needs to provide incentives to promote innovative and environmentally friendly food handling, preservation, and value addition technologies.

Market and traders' associations are important governance nodes in the urban food system. They impact the access, affordability, and quality of food. Most of these are established to lobby for the welfare of traders and improve access to food; however, because they control entry, allocate space to their members, and sometimes dictate food prices in the markets, they have been branded as cartels. Middlemen also play a significant role in the food system of Kisumu, facilitating collection of food from production sources and its distribution in the city, linking farmers to markets. However, they have also been demonized for dictating food prices in the markets.

Financing and access to capital is another challenge to the food retail sector in Kisumu, which is largely driven by small-scale informal traders. Capital is evidently hard to come by as 63% of the traders surveyed have never had any form of financial support for their businesses. For those who have received financial support, the most common form of support came from savings club members (16%) and family members (8%). These small-scale, underfinanced food retailers are further burdened by several taxes and operating fees, including inspection fees, food handlers license fee, market fees, and in some instances protection fees to market cartels and criminal gangs. These costs are passed on to the consumer, making food more expensive.

The changing urban food economy, particularly the entry of large supermarkets chains, also poses challenges to small-scale food retailers. Supermarkets are now offering goods that were traditionally sold in municipal markets and kiosks, e.g. fruits and vegetables; and they are also moving into residential areas, pushing small-scale informal traders out of business. With rising unemployment, many more people are also getting into food retail business as a source of livelihood, leading to heightened competition even among the informal traders. In the CUP food retail survey, 33% of traders mentioned competition from other traders as a key threat to their businesses; however, while competition is a threat to the traders, it is good for consumers who benefit from competitive prices.

Conclusion

The urban food system plays an important role in the economy of Kisumu. Food retail is a source of employment and livelihood for many, particularly women, who traditionally are engaged in food provisioning, preparation, and marketing. Informal food retailers dominate the sector, enabling access to food by the poor. The urban food system is premised on a supply system that is dependent on distant production locations, which in turn introduces complex challenges in food supply including infrastructure, seasonality, middlemen and cartels, and price escalations. The current governance of food retail businesses for informal sector traders has been shown to fail to address the issues of infrastructure and service provision that would enhance food safety, lower supply prices, and improve preservation.

References

FAO, (2016). *Food for cities programme: Building sustainable and resilient city region food systems.* Available at: www.fao.org/in-action/food-for-cities-programme [Accessed 12 Jan. 2017].

IIED, (2016). *Informal food systems and food security in rural and urban East Africa.* Available at: http://pubs.iied.org/17336IIED [Accessed 23 Feb. 2017].

Mireri, C., Atekyerezab, P., Kyessic, A. and Mushic, N. (2007). Environmental risks of urban agriculture in the Lake Victoria drainage basin: A case of Kisumu Municipality, Kenya. *Habitat International*, 31, pp. 375–386.

Omollo, K. (2015). Construction of 3 hawkers' markets begins in Kisumu. *Standard Digital*. Available at: www.standardmedia.co.ke/article/2000174190/construction-of-3-hawkers-markets-begin-in-kisumu [Accessed 26 Aug. 2017].

Onyango, J.O., Olima, W.L.A. and Onyango, L. (2012). Dynamics of street vending phenomenon in the Kisumu Municipality, Kenya. *International Journal of Arts and Commerce*, 1(4), pp. 107–120.

Skinner, C. and Haysom, G. (2016). The informal sector's role in food security: A missing link in policy debates? *PLAAS, Working Paper 44*. Cape Town: University of the Western Cape.

UN-HABITAT, (2005). *Situation analysis of informal settlements in Kisumu*. Cities without Slums Sub-Regional Programme for Eastern and Southern Africa. Nairobi: United Nations Human Settlements Programme.

14 The characteristics of the urban food system in Kitwe, Zambia

A focus on the retail sector

Issahaka Fuseini, Jane Battersby, and Niraj Jain

Introduction

The Consuming Urban Poverty (CUP) project as a whole is premised on the assertion that the food system is a useful lens through which to understand and alleviate urban poverty. In the introduction to her book *Hungry City*, Carolyn Steel (2008) argues that food and cities are inexorably linked, and that it is important to view the cities as much shaped by food, as the food system is shaped by cities, as she notes: "Every day we inhabit spaces food has made" (Steel 2008: 1). As such, it is therefore important to consider how food and cities intersect to create conditions that shape urban poverty.

This chapter presents some of the findings of the study of Kitwe's food system, using food retail as an entry point to address three central considerations. Firstly, what do the characteristics of the food retail sector reveal about the nature of Kitwe's economy and employment opportunities? Secondly, what do the characteristics of the food retail sector reveal about the lived experience of poverty in Kitwe? And finally, how do issues of infrastructure and other challenges shape the food retail sector, and therefore impact poverty and livelihoods?

The development of Kitwe's food system

Background on the economic development of Kitwe and its size, growth, and spatial development is contained in the Introduction to the book. The development of Kitwe's food system has been shaped by and has shaped the economic, political, and social development of Kitwe city. In the early years of Kitwe's development, the city's food system was dominated by paternalistic practices of mining companies providing food rations to their workforce (Parpart 1986; Dandule 2012). It was deemed important to integrate the provision of food into the management of African mine workers as many of the new recruits were often found to be underweight. The food rations were therefore strictly controlled to provide just enough caloric requirement for work and in addition to control the influx of migrants from rural areas, particularly the spouses of mine workers. As such this policy included restrictions on married mine workers from settling in the mine compounds with their spouses, choosing instead to grant married workers annual leave to visit their families in the hinterlands.

While maintaining the food rations for mine workers, the decision to allow women and wives into the mining compounds led to increased demand for food. This served as the watershed moment when other components of the food systems of Kitwe and other mining towns in the Copperbelt were impacted in terms of production and product mix. For example, efforts to feed both mine workers and their families necessitated the colonial authorities to promote commercial maize production, introduce the cultivation of cassava, and even allow and grant land to the accompanying women to engage in home gardening to supplement their households' rations (Dandule 2012). The increased demand for food from urban settlements across the Copperbelt Region provided a secure market for commercial maize and beef farmers. As discussed in Chapter 17, this historical evolution of the food system in Zambian mines partly explains why the contemporary food system is dominated by maize. In Zambia, maize has come to symbolise food security or even the food system, and as such had acquired a high political status, with successive governments intervening to make the product available and accessible through production and marketing subsidies.

Following consistent decline in the 1990s, Zambian mines were privatised, which resulted in a large-scale retrenchment of mine workers (Hampwaye 2008), and therefore losses of household income, cancellation of provision of food by mining companies, and a liberalised economy that enabled the greater exposure of the cities' food system to global forces (Fraser and Lungu 2007; Mwitwa and Ng'andwe 2010). This profoundly reshaped the food system, shifting both demand and supply. One of the most apparent shifts was the emergence of small-scale and informal food retail as both a means to supply residents no longer fed by the mining companies, and as a livelihood for retrenched workers and their families no longer supported by mining salaries. Thus, the current food system of Kitwe, particularly the food retail component, emerged from historical economic restructuring processes as the city's residents sought to cope with the changing economic realities regarding employment, livelihoods, and their food needs. While the focus of this chapter is on the contemporary food system, it is clear that the characteristics of this food system remain shaped by Kitwe's history.

Research methods

The CUP project used a range of research methods. The work presented here draws on the largely quantitative food retail survey conducted as part of the wider project. Respondents were selected through a two-stage sampling procedure. First, a retail mapping exercise was conducted in sites across the city. These were Chisokone (the city's main central market), the Central Business District, the neighbourhoods of Ipusukilo, Wusakile, and Chamboli, and sites around shopping malls. This retail mapping produced a total of 705 food retail outlets. The second substantive sampling procedure involved working out a representative sample from the 705 food retail outlets. This resulted in 375 retail outlets in the same sites being selected for the actual retail survey. The survey, conducted on tablets using Open Data Kit (ODK) and uploaded onto the Ona data platform, addressed issues of business structure and practices, foods sold and the sources of these foods, access to infrastructure, and challenges experienced by businesses inter alia.

Poverty, livelihoods, and characteristics of food retail in Kitwe

The privatisation of the mines in Kitwe dramatically reduced levels of formal employment and thus encouraged the movement of ex-miners and their family members into the informal sector to meet their livelihood needs (Fraser and Lungu 2007; Hampwaye 2008). In the years since this period of retrenchment, the economy of Kitwe has become increasingly dominated by service sector employment. The data generated by the CUP survey are not longitudinal, but the data can be used to reflect on current economic conditions as shaped by the city's history.

Businesses mainly identified as permanent (n = 310) rather than cyclical (n = 26) or seasonal (n =17), and the data on the years of operation indicated that many of the businesses had been operating for a considerable length of time. Although often viewed as a transitory stopgap activity, 41% of food retail businesses had been in operation for six years or more, and a further 26% for three to five years. This suggests that food retailing provides a relatively stable livelihood; however, it was also demonstrated to be a fairly marginal livelihood. Of the surveyed retailers, 46% said that more than half of their household income was derived from the store, but 81% of retailers indicated that this income was sometimes insufficient to meet household needs and a further 7% indicated that it was always insufficient. This therefore suggests that the food retail sector is a relatively marginal business opportunity, but in the context of few alternatives, one that is sustainable for participants. The relatively marginal nature of these businesses was further evident in the fact that very few storeowners operated more than one business (just 23 traders out of 389) or had permanent or casual workers working for them.

As Table 14.1 illustrates there was diversity within gender and age of retailers. The majority of retailers (58%) were female, and 88% were between 18 and 50 years old. The age distribution suggests that many of the food retailers spend their working age in the business. From Table 14.1 it appears that males tend to enter the food retail business earlier in their lives and also exit earlier than females. Forty-six percent (53 out of 121) of male food retailers were 35 years old or younger, compared to 32% of female traders. This suggests either that male engagement in the food retail is more transient than women's engagement, or that young males are increasingly turning to the food retail sector in the absence of employment opportunities within other sectors. The characteristics of food retail sector workers in terms of gender, permanence or otherwise, and length of time in business provide important insights into the structural transformation of the city's economy.

Table 14.1 Age–gender cross-tabulation of food retailers in Kitwe (N=284)

	Gender of respondent					
Age	*Female*		*Male*		*Total*	
	Freq.	*%*	*Freq.*	*%*	*Freq.*	*%*
16–17	2	1	3	1	5	2
18–35	50	18	53	19	103	36
36–50	92	32	56	20	148	52
51–60	17	6	8	3	25	89
61+	2	1	1	0	3	1
Total	163	57	121	43	284	100

The research revealed the nature of interaction between food retail and Kitwe's economy and employment opportunities. The characteristics of food retail are indicative of the historical structural dynamics of the city's economy as well as employment and livelihood pursuits of its inhabitants. The decline in the mining sector, the loss of free food rations to mine workers, and a liberalised economy has meant that food retail emerged to fill important gaps created by these processes. As illustrated in the following sections, the changes have reoriented the economic activities in ways that reflect a degree of community solidarity, sometimes defying classical economic logic of profit making.

Food retail and the urban consumer

The business practices of the food retailers are responsive to the social and economic dynamics of Kitwe's residents. The sector adopts strategies in marketing (bulk breaking, offering credit, operating times, restocking, etc.) to respond to the needs of different segments of the city's population. These strategies are discussed in the paragraphs below.

Bulk breaking of food products

This involves retailers breaking their wares into smaller and affordable units that match the incomes or demand of their clients. The practice of bulk breaking is seen as a strategy aimed at improving access to food by and for the poor segment of society. The bulk-broken products may have slightly higher per unit costs, but are still affordable to many a poor urban household that depends on daily, casual, or unstable incomes (Battersby 2012).

As Table 14.2 indicates, bulk breaking is widely practiced. Dairy products, sugared drinks, and bread were less likely to be broken and sold in smaller units compared to the other food products. This is likely because these food products are originally packaged in quantities with unit prices that are within the economic reach of the poor households in the city. Battersby (2012) argues that bulk breaking

Table 14.2 Incidence of bulk breaking of selected food products (based on reported sales of those products)

	Yes	*%*	*No*	*%*	*Total*	*%*
Vegetables	57	97	2	3	59	100
Fruits	20	95	1	5	21	100
Roots and tubers	29	97	1	3	30	100
Dry grains	45	92	4	8	49	100
Dry pulses	25	89	3	11	28	100
Fresh meat	6	100	0	0	6	100
Frozen meat	11	92	1	8	12	100
Fresh fish	12	100	0	0	12	100
Dried fish	23	82	5	18	28	100
Frozen fish	15	100	0	0	15	100
Smoked fish	8	89	1	11	9	100
Eggs	96	82	21	18	117	100
Dairy	53	63	31	37	84	100
Oils and fats	83	86	14	14	97	100
Sugared drinks	94	69	43	31	137	100
Sweets	103	93	8	7	111	100
Sugar	100	81	24	19	124	100
Bread	68	63	40	37	108	100
Groundnuts	26	93	2	7	28	100

is driven by the mismatch between the socio-economic realities of poor households (demand side) and food retail strategies based on minimum unit size and associated pricing regime (supply side). According to Battersby, while income is an obvious undercurrent of the practice of bulk breaking, other factors such as lack of adequate storage space and limited refrigeration facilities add to and reinforce the need to retail food products in smaller units commensurate with the incomes and infrastructural realities of households. The responsiveness of retailers to the purchasing capacity/demand of consumers is one reason for the high patronage of such retailers by households in Kitwe, as will be discussed in Chapter 17.

Operating times

A second form of responsiveness to consumer needs is long operating hours. The nature of urban employment for poor residents often means that workers leave early for work and return late. The food retail sector in Kitwe is responsive to this lived reality. About 31% of retailers opened between 4am and 6am, providing opportunities for consumers to access food before leaving for work or en route to work. At the other end of the day, 53% of retailers closed only after 7pm, thus enabling people to access food after the working day. Supermarket opening hours are more limited and their locations less physically accessible, being located only in central city areas. The trading hours noted above are typical of the informal food retail especially with respect to neighbourhood and home-based shops. It is therefore likely that the frequency with which households patronise neighbourhood and home-based outlets is not driven solely by geography, income, and bulk-breaking practices but may also be the result of suitable operating hours by those food retail outlets. Extended, flexible operating times bring retailers and customers together in a mutually beneficial manner (Peyton *et al.* 2015).

Offer of credit

A final way in which food retail impacts on the city's food system and household access to food is the habit of granting credit by retailers. This practice is common with the informal retailers (Battersby 2012; Peyton *et al.* 2015; Skinner and Haysom 2016). Just under a third of retailers surveyed granted credit, with 27% offering credit regularly and 3% granting it "sometimes". Asked who would be granted credit, the respondents ranked regular customers (38%) above all others on their list of likely beneficiaries of credit overtures, followed by clients with "good credit". Clients with good credit are those who have a history of honouring their credit obligation with the creditors.

The practice of granting credit was based on responsiveness to and reflective of the conditions of poverty in Kitwe. Of those granting credit, 38% said they offer credit to retain customers. Beyond the business logic of responding to consumer economic need, about 30% offered credit on social grounds claiming it is part of their culture to do so, that it is necessary in a poor community, and that without credit overtures to some societal members many would go hungry. A further 26% couched their rationale for granting credit as a business strategy to attract consumers. Therefore, the practice of offering credit to clients is not only a business strategy to retain and attract customers but also a safety net that poor urban households draw on to mitigate any exposure to food insecurity or despondency. This relationship is premised on the socially binding obligation to help others in times of crisis with the expectation that the beneficiary might also reciprocate the gesture directly or indirectly to other members of society who might also be in need.

The social principle underlying credit is further illustrated by the fact that about 93% of retailers who offer credit do not charge interest, so that beneficiaries are genuinely assisted to overcome their economic difficulties. However, there were mixed results in terms of repayment of credit. About 61% of the respondents considered repayment rate as satisfactory, 25% indicated a good repayment rate, and 36% reported average rates of repayment. On the other hand, 39% of the respondents thought repayment of credit was problematic. Despite this, the losses of income due to bad debt were not high. Eighty-two percent of the retailers who grant credit reported not losing income of up to 5% due to bad debt. This means that despite the high repayment challenge, retailers manage their credit portfolio in ways that minimise losses, and that is important to keep the underlying mutual relationship going.

Through this basic analysis of business practices of food retailers in Kitwe it is clear that their activities both reflect and respond to lived poverty in the city. Put differently, the functional characteristics of food retail in Kitwe are the reflection of experiences of lived poverty among poor residents of the city. The marketing strategies used by retailers – bulk breaking, flexible operating times, and offering of credit – are the results of demand–supply interactions underpinned by the lived poverty experiences of the urban poor.

Product mix of food retail in Kitwe

As illustrated in the previous section, Kitwe's food retail sector is responsive to the lived experience of the city's poor residents. This facilitates basic access to food. However, it is important to consider what kind of foods are being made available within this system, and how these connect to the conditions of poverty in which both retailers and residents operate.

As Figure 14.1 indicates, the food retail environment in Kitwe is characterised by a wide range of food products, but carbohydrate-dense and sugary foods are the most commonly stocked foods. Except for eggs, protein-rich foods are generally low in representation as are pulses and fruits. Also noteworthy is the fact that processed foods are more prevalent than unprocessed foods. The high incidence of "tea and coffee" and sugary products reflects the intersection of Kitwe's food system with its urban economic history. Parpart (1986) reports that during the 1940s European foods such as tea, sugar, and bread became necessary staples in Kitwe as mine workers tried to adopt "proper" urban lifestyles.

The balance of foods sold is shaped by the multi-dimensional nature of the poverty of both consumers and retailers. Relative prices of food mean that poor urban residents often have diets that are calorie dense and nutritionally inadequate, as reflected in the product mix here. In addition to the relative costs of foods, these choices are shaped by factors such as limited refrigeration and storage capacity which then determine volumes of food purchased and ensure that processed foods that are less prone to spoilage are more commonly consumed. Likewise, issues of storage and refrigeration as infrastructural challenges shape the types of foods stocked and sold by the retail sector. Additionally, the high presence of snack foods is perhaps an indicator of relative poverty. These are items that can be bought at low cost with limited transportation requirements and are therefore viable entry points into the food retail sector for the poor.

The following section considers questions of infrastructure and identifies barriers to trade in order to reflect on potential entry points for policy makers and planners that could improve food retail in Kitwe to alleviate poverty for retailers and consumers.

Access to enabling infrastructure

Infrastructure deficits impact the safety and viability of food retail. In this regard, access to both physical (shelter, storage, and refrigeration) and service (water, sanitation, and electricity) infrastructure is very important. Figure 14.2 shows the retailers' access to key infrastructure relevant to food retail.

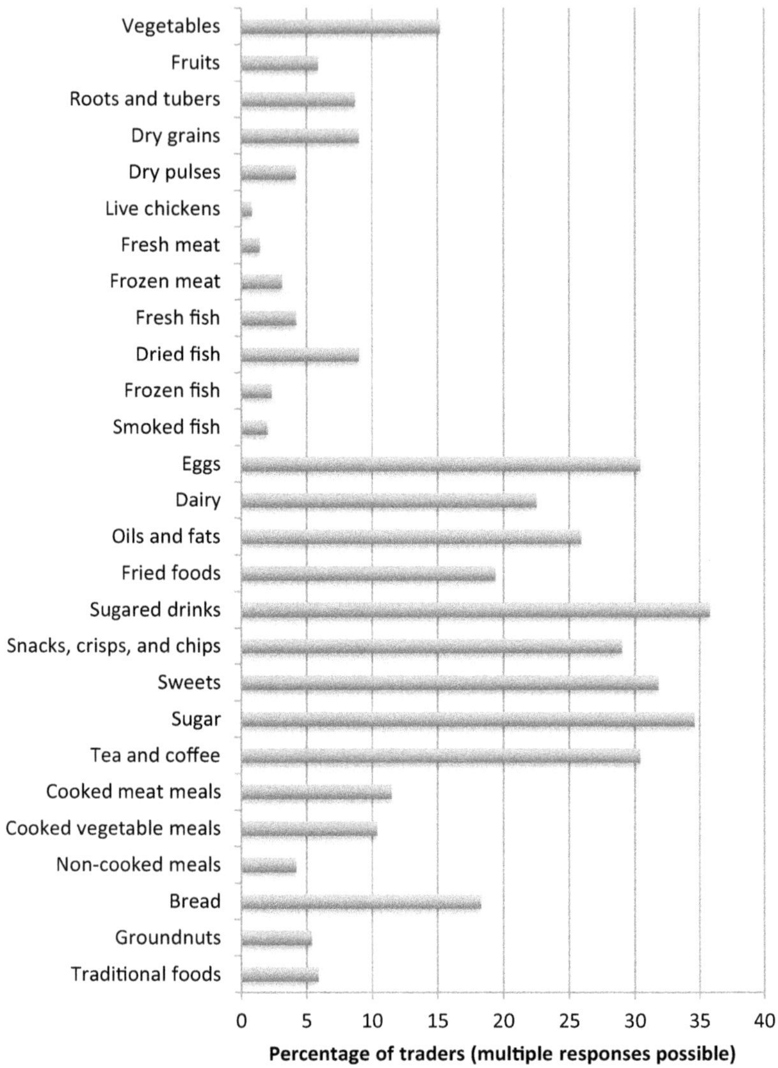

Figure 14.1 Food products sold in Kitwe by retailers (% of cases)

(Author's own)

Although just over half of retailers had access to waste disposal facilities (52%), significantly fewer had access to piped running water (23%) or sanitation facilities (33%). For retailers wishing to sell fresh produce, meat/fish, or prepared foods, this infrastructure deficit significantly impacts potential food safety. This may account for the relatively high proportion of retailers choosing to sell processed foods. This then potentially increases

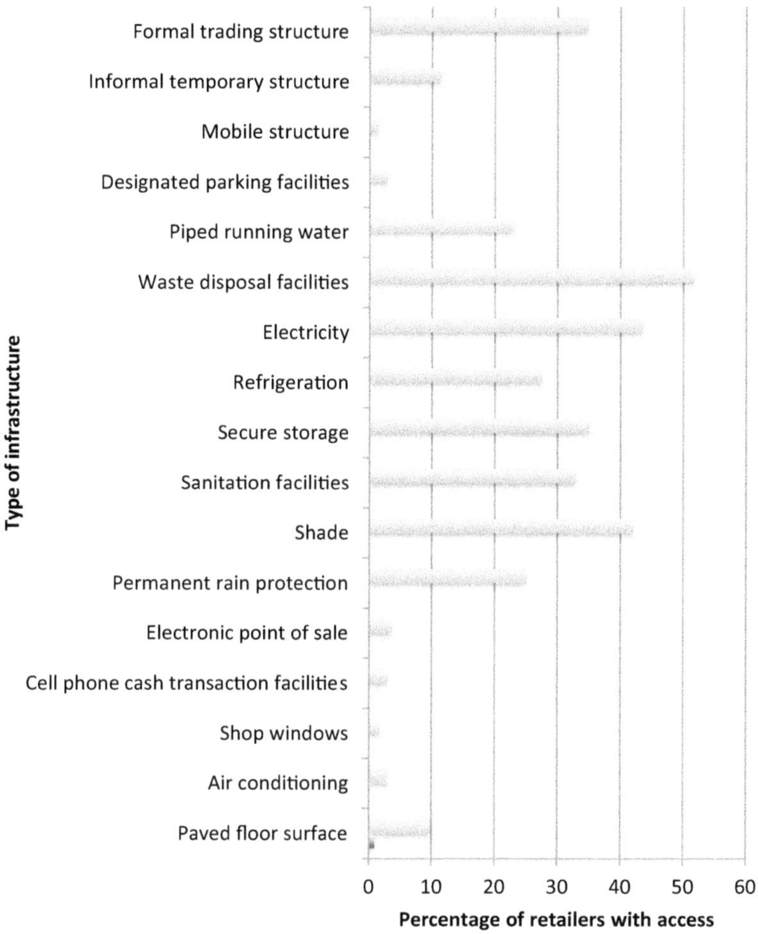

Figure 14.2 Access to supportive infrastructure by food retailers in Kitwe

(Author's own)

the burden of disease for the low-income residents who are dependent on these retailers for food.

A second set of infrastructural deficiencies was refrigeration and secure storage (accessible to just over a quarter and just over a third of retailers, respectively). Refrigeration and storage are important determinants of retail stocking and operating strategies, shaping what foods are stocked and how often foods are procured. The limited access to appropriate infrastructure has led to retailers adopting operating strategies to serve their clients regularly and minimise losses through spoilage. This strategy is embodied in frequent restocking of their wares, as it allows traders to overcome the problem of limited storage space and refrigeration facilities. As illustrated in Figure 14.3, perishables or food products with short shelf lives (such as vegetables, fruits, freshly prepared foods and bread) had the highest frequencies of restocking, and products with longer shelf lives (e.g. roots and tubers, dry grains, sugar, sugared drinks, oils and fats, sweets) were restocked less frequently. With these latter products, weekly and monthly restocking were more common, with annual restocking in some cases.

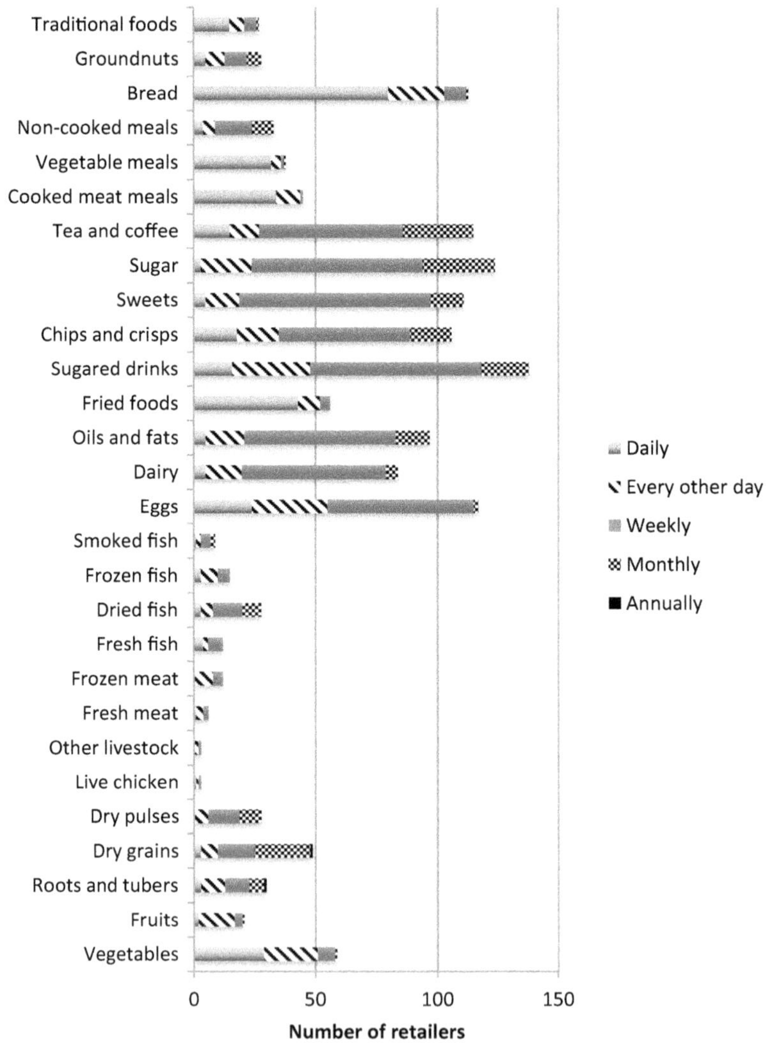

Figure 14.3 Frequency of restocking food products by retailers

(Author's own)

This frequent restocking of products with short shelf lives was a response to both the infrastructure deficits of retailers and of consumers. This is where business logic meets with customer economic and social realities. Many poor urban households suffer a similar lack of infrastructure to retailers (e.g. refrigeration) and thus have to buy perishable and fresh products on a daily basis (Skinner and Haysom 2016). This strategy may be very useful for the many informal food retailers who lack bare infrastructure (e.g. street venders) and thus operate mobile outlets. Even though frequent restocking has potential operational costs, such as transport,[1] it nonetheless sustains retailer–client relationships through regular product availability and quality. As we have seen in the cases of bulk breaking, flexible and extended operating times, and the habit of offering credit to clients, the practice of frequent restocking is another strategy geared at building the trust which is necessary for a long-term positive retailer–client relationship.

Challenges of food retail in Kitwe

In addition to the infrastructural issues highlighted above, there are other substantive challenges facing food retail in Kitwe. Retailers were asked to reflect on threats to their businesses (Figure 14.4). Competition from other stores

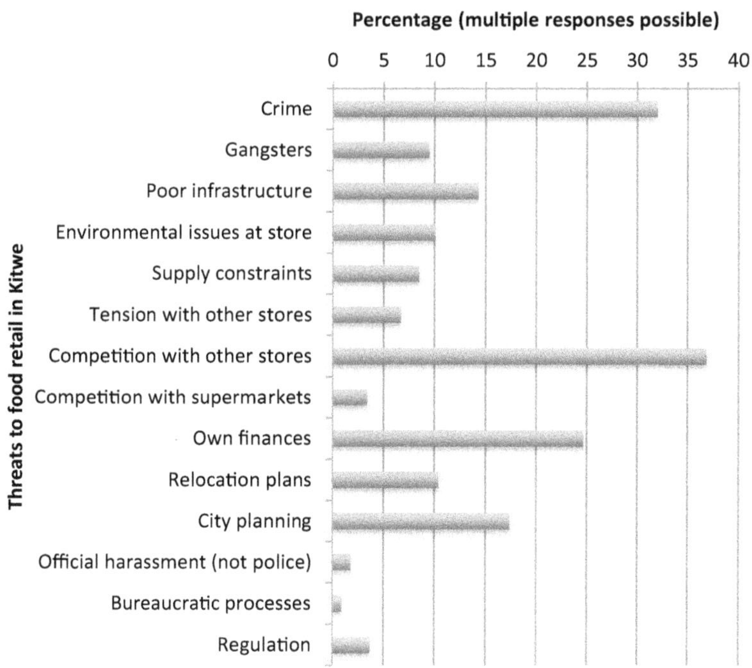

Figure 14.4 Respondents' views on major threats to their food retail businesses

(Author's own)

(37%), crime (32%), own finances (25%), and city planning (17%) were identified as the other most important threats to the sustainability of their food retail businesses. Food retail in Kitwe is characterised by involution (Maxwell 1998), in which many small businesses, employing a small number of people, sell the same products in the same sites. This characteristic explains the concerns about the competition from other stores. Concerns about crime also rank relatively highly. The reasons for this were unfortunately not interrogated further, but municipal market managers expressed concerns in person about the density of retail enabling the sale of illicit goods, such as alcohol. This could at least partially explain high crime rates. Theft of stock is also viewed as a challenge, and one that could be mitigated by better access to storage.

The limited finances of retailers, as a manifestation of poverty, was identified as a critical challenge. This shaped business practices, such as stocking practices, ability to source products directly and avoid transportation middlemen, and to offset periods of food price increases or decreased demand (Siyanga 2017). The relatively high concern about city planning reflects the ongoing tensions between market traders and Kitwe City Council regarding the upgrading and replacing of Chisokone Market, as described in Chapter 9.

Exclusionary city planning adds to the challenges faced by food retail, as described in Chapter 10. The informal retail sector largely operates in contravention of planning codes. This illegality reinforces the limited access to key infrastructure and ultimately impacts negatively on the operations of those businesses. The other challenges identified in Figure 14.4 intersect to constrain the operations of food retail in the city. In sum, the infrastructural deficits interact with other fundamental constraints to shape the nature of food retail in Kitwe, and ultimately impact negatively on the livelihood activities and poverty experiences of the poor urban households.

Conclusion

The current food retail sector in Kitwe is informed by the historical economic conditions of the city through its growth as a mining town, its subsequent economic decline, and its re-emergence as an economy centred on the service sector. The nature of employment, livelihoods, and business characteristics illustrate marginal, but sustainable, businesses in which the food sector offers employment for women with few barriers to entry, but few opportunities for growth.

The chapter argued that the practices of food retailers in Kitwe, with regards to what foods they stock, bulk breaking, extension of credit, and opening hours, provide insights into both the lived experience of poverty of Kitwe residents making demands of a deeply embedded local food retail that needs to be responsive to consumer needs, and the poverty experienced by retailers in the forms of financial and infrastructure deficits. The chapter therefore illustrates that poverty and the food system are connected in a number of ways through supply and demand issues. This therefore presents opportunities for policy makers to engage the food sector as a viable entry point to alleviate poverty.

Food retail in Kitwe has been found to be providing access to food through a means of business practices informed by the lived experience of Kitwe's residents. However, it was also found that the mix of products sold is not supportive of healthy diets. This chapter has argued that this is the result of infrastructure deficiencies for both retailers and consumers. Provision of appropriate infrastructure (water, sanitation, storage, refrigeration) should be integral to doing policy framing with respect to the above point of creating supportive structures for food retail as a livelihood activity. This could improve the health and safety of food retail, enable retailers to stock more nutritious foods, and purchase in greater volumes, thereby reducing cost to the consumer. To a large extent, the reputation of the informal food retail as unsafe is a function of access to requisite infrastructure (Skinner and Haysom 2016). This chapter ultimately argues that the existing role of the food retail sector of responding to the food needs of the poor could be enhanced by greater attention by policy makers to helping retailers overcome barriers to selling affordable and nutritious foods.

Note

1 Over 82% of the surveyed retailers indicated that the cost of transportation was the most important cost they incurred other than stock. The frequent restocking could be a major contributor to the cost of transportation.

References

Battersby, J. (2012). Beyond the food desert: Finding ways to speak about food security in South Africa. *Geografiska Annaler, Series B: Human Geography*, 94(2), pp. 141–159.

Dandule, B. (2012). *Women and mineworkers' struggles on the Zambian Copperbelt, 1926–1964*. Master, University of Zambia.

Fraser, A. and Lungu, J. (2007). *For whom the windfalls? Winners and losers in the privatisation of Zambia's copper mines*. Lusaka: Civil Society Trade Network of Zambia and Catholic Commission for Justice, Development and Peace.

Hampwaye, G. (2008). *Decentralisation, local economic development and urban agriculture in Zambia*. PhD, University of the Witwatersrand.

Maxwell, D.G. 1998. The political economy of urban food security in sub-Saharan Africa. *Journal World Development*, 27(11), pp. 1939–1953.

Mwitwa, J. and Ng'andwe, P. (2010). *Effects of the global financial crisis on the food security of poor urban households: Case study Kite, Zambia*. Kitwe: RUAF and Copperbelt University.

Parpart, J.L. (1986). The household and the mine shaft: Gender and class struggles on the Zambian Copperbelt. *Journal of Southern African Studies*, 13(1), pp. 36–56.

Peyton, S., Moseley, W. and Battersby, J. (2015). Implications of supermarket expansion on urban food security in Cape Town, South Africa. *African Geographical Review*, 34(1), pp. 36–54.

Siyanga, F. (2017). *From farm to fork: Understanding Kitwe's food system through the fish lens*. Unpublished MPhil Thesis, University of Cape Town.

Skinner, C. and Haysom, G. (2016). The informal sector's role in food security: A missing link in policy debates? *Working Paper 44*. Cape Town: PLAAS, UWC, and Centre of Excellence for Food Security.

Steel, C. (2008). *Hungry city: How food shapes our lives*. London: Chatto & Windus.

15 The characteristics of the urban food system in Epworth, Zimbabwe

Godfrey Tawodzera, Easther Chigumira, Idah Mbengo, and Samuel Kusangaya

Introduction

Food security is fundamental to promoting development (FAO 2017a). In sub-Saharan Africa, in the face of high levels of poverty and stagnating economies, food insecurity poses a serious challenge to development (Chauvin *et al.* 2012). Given the nature of urbanisation and the scale of urban poverty articulated in Chapter 1 of this volume, it is likely that food security will remain a development concern in African cities for the foreseeable future, particularly in secondary cities, where much of the urban growth in Africa is occurring. Tacoli (2016) argues that food can be the entry point for a more holistic understanding of urban poverty that includes food security (and safety) and also access to decent livelihoods. This therefore requires a broad and nuanced understanding of the entire urban food system. As Chapter 3 notes, the failure of the current remedial urban food security approaches stem from an inappropriate focus on the household level which neglects the wider systemic drivers of food insecurity at the broader urban food system scale. Some of these drivers are rooted in the ways in which the food system is governed both at national and sub-national scales. Laws made at the national level, for example, influence how trade occurs at the local scale and how households access food. Chapter 10 provides detailed information on laws affecting the food system in Zimbabwe through a discussion of the country's political economy and the key events that have affected the food system, including the Economic Structural Adjustment Programme (ESAP), the Fast Track Land Reform Programme, Operation Murambatsvina, dollarization and various statutory instruments (e.g. SI 64 of 2016). In order to address food poverty, it is therefore essential to consider how this poverty is shaped by conditions within the food system. Fundamental to this consideration in Epworth is the centrality of the informal sector in shaping the food system of the area. This chapter will thus address two questions: a) What are the major food sources for residents in Epworth? and b) How do informal food retailers shape Epworth's food system?

Research methods

In order to address the key issues identified above, three research components were carried out in Epworth in 2016: a) a retail mapping census; b) a retail survey; and c) a reverse value chain analysis of five key items consumed by the poor. The retail mapping census enumerated and mapped all 1607 food retail stores that were identified in the area, in the process collecting data on food retail outlet distribution, store typology and foods sold. This information was necessary to understand Epworth's food system, especially which areas were better served and which retail types were dominating in different locations. The location of retail food outlets impacts on residents' food *access*, which is a key component of urban food security. The retail survey involved carrying out an in-depth enquiry into retail outlet operations with a statistically determined representative sample (in terms of gender, typology, distribution, and product) of 293 retail outlets from the total of 1607 retail outlets identified in the retail mapping exercise. The five-item reverse value chain analysis identified five key food items (maize-meal, rice, *maputi*, vegetables, and offal) in Epworth and tracked these foods back along the value chain, identifying the multiple packaging and pricing processes that occurred along the value chain, stopping either at the farm, lake or international source, or place where no further information was available. This reverse approach increased the chances of identifying more suppliers to the area than would have been the case in a traditional "farm to fork" value chain analysis, which ordinarily pre-selects well-known suppliers and tracks them forward into the consumer arena, often missing new, lesser known and unconventional supply chain actors and even whole value chains. Together, the three different but complimentary research methods produced novel data on Epworth's food retail system and helped to identify the three aforementioned aspects of Epworth's food retail system.

Study results

This section presents and analyzes findings from the Epworth's food system study. Most modern urban food systems are generally characterized by modernized wholesale systems, capital-intensive food processing, state-of-the-art logistics, private branding, labelling (and packaging) modern retail and restaurants, and global integration (Tefft *et al.* 2017). While Epworth's food system is undoubtedly located within, and is influenced by, macro-level operations, the food-related processes at the local level are powerful and these are predominantly operating within the realm of the informal sector. Although informal food systems have traditionally been ignored or penalized by governments (Tacoli 2016), the findings of this study indicate the crucial role that the informal food sector plays in shaping the food systems of urban areas such as Epworth.

Sources of food in Epworth

Food sources are integral to food systems. The viability of a food system is generally measured by how it enables or impedes access. While in urban areas accessibility hinges primarily on the individual's or household's ability to purchase foodstuffs (Crush and Frayne 2011), another component of food access is also contingent upon where the food is sourced and retailed. Swift and Hamilton (2001) argue that retail distribution systems are generally tilted in favour of high-income areas where infrastructure and marketing systems are well established and functional. In contrast, network systems in low-income areas are poor and food reaches low-income areas through informal rather than formal systems (Battersby and Crush 2014). The reverse value chain analysis carried out in this study served to understand not only where the food comes from, but also the changes in pricing that occur along the value chain to impact on economic access and ultimately on food poverty.

The value chain of each of these items was reviewed from the point of retail to the source, or as close to the source as possible. The value chain review identified a number of general trends. One of the trends is that value chains are fluid, with food being sourced from a number of different places. These sources included international suppliers, often from within the southern African region. The value chain for maize-meal was particularly informing: most of the maize-meal that was retailed in Epworth was imported from South Africa and Zambia. While the government also imported maize-meal through the Grain Marketing Board (GMB), its role was minimal. The greater proportion of the maize-meal was therefore imported by cross-border traders who then sold it to wholesalers.

Most wholesalers in Harare were generally wary of being involved in direct importation because of the cost and bureaucracy surrounding importation permits. Cross-border traders, on the other hand, were able to bring in small quantities that did not require permits or payment of duty, making numerous trips per month to keep wholesalers adequately supplied. The contribution of cross-border traders has therefore been significant, underlying the key role that the informal food sector has been playing in the food system of the country in general, and for Epworth in particular. Small traders in Epworth also bought maize-meal from wholesalers in Harare and repackaged it into smaller quantities that were required by most of the consumers whose low incomes could not allow them to purchase in bulk. In addition, the role of the informal food sector underlined the innovative ways that poor consumers in Epworth were responding to food-related challenges: consumers were reverting to purchasing whole maize and milling in place of ready packaged maize-meal from retail stores. Most of these maize-meal purchases are made from informal traders, from the local grinding mills or *chigayos* or from the market at Mbare Musika in Harare. This has a moderating effect on the prices as residents pay an average of US$7 (Sept. 2017) for a 20-kg bag of milled mealie-meal instead of the US$10 it is sold for in retail shops. The small maize-meal packages sold by vendors could be as low as 1kg or 2kg and priced at US$0.50 and US$1 respectively, hence enabling the poor to access food using the little money they would have raised.

The rice value chain supply into Epworth was generally long: while some of the rice came from South Africa and Mozambique, some of the rice was coming from China, Vietnam and Singapore. The length of the rice value chain reaffirms the embeddedness of Epworth's food system into the global and regional food system. The major players in the importation of rice were wholesalers, the government and cross-border traders, indicating the multiple entry points for rice into the country's food system. Zimbabwe imports close to 95% of its rice requirements and the importation regulations are not as stringent as that of maize. The rice imported by government, through the GMB, was generally in 50-kg bags, which were repackaged into smaller ones under the "Silo Brands" label. The majority of traders purchased rice from wholesalers in Harare where a 20-kg bag of rice cost, on average, US$18.50. A 2-kg packet of rice would then be traded for US$2.00 in retail shops and by street vendors. Par-boiled rice from South Africa was cheaper (US$1.80 per 2kg) than other rice varieties on the market. This was primarily because of the shorter value chain for this commodity. The rice coming in from Mozambique, however, did not follow this trend as the value chain was discovered to be much longer, with the rice being originally from Asia. This finding validates Gitz's (2016) argument that short value chains are beneficial to the food system through their impact on reducing prices. With shorter value chains, Epworth is likely to reap the benefits of reduced food prices, which could potentially reduce food poverty in the area.

The problem of longer value chains and their detrimental effect on food pricing is seen in Epworth as foods are often retailed at a higher price, largely as a result of the cost of transport, even when they came from central Harare. Foods that are important to poor consumers, particularly rice, are not seen as being important by policymakers and, as a result, different actors within the value chain are able to assert disproportionate power along the value chain, resulting in price volatility and product instability.

Another important food in Epworth is vegetables, particularly *covo* and *rape* (kale-type leafy greens) which are generally part of the residents' dishes. Although residents grow vegetables in Epworth, the majority grow for own consumption and not for sale. Vegetables traded in the area are mainly bought from Mbare Musika in Harare as Epworth does not have a market for farmers to deliver. Mbare Musika is supplied by farmers from nearby Mazowe, Dema and Marondera, but also from as far away as Mutoko, some 140km from the capital. Thus, the vegetable trade, which is an important component of the food system in Epworth, is driven externally from Mbare Musika, where pricing is determined. Major players in the determination of vegetable prices are the middlemen (*makoronyera*), who control access for farmers and buyers. A bundle of vegetables at Mbare Musika costs, on average, between US$1.50 and US$2.00 depending on demand. In times of short supply, the cost can rise to US$3.00 per bundle. A round trip from Epworth to Mbare Musika may cost a trader US$4.00, adding significantly to costs incurred by these traders. These costs are later passed on to consumers, who have to pay US$0.10 for five to six vegetable leaves. In comparison, residents in Mbare pay the same amount for double the amount of leaves. Thus, Epworth residents pay significantly higher prices owing to the way the value chain for vegetables is structured. In addition, the value chain for vegetables predominantly operates in the informal food sector, with the trade involving street vendors, mobile operators with push-carts, wheelbarrows and those trading on bicycles.

Amidst all the food challenges in Epworth, an "innovation" has arisen in the area in the form of a snack food called *maputi*, a type of popped corn. Maputi is rapidly gaining popularity in Epworth and is resulting in interesting value chain related responses (see Muramba, forthcoming). Most households in Epworth cannot afford three solid meals per day. To bridge between meals, most household members resort to consuming maputi. This relatively affordable product is rapidly replacing more traditional street foods (such as roasted maize). This is because its price has virtually remained the same for close to three years. Traditionally, most of the maputi sold in Epworth used to be processed in Mbare, Msasa Industrial Park and Chitungwiza. Of late, however, maputi is being locally produced in Epworth as many small-scale manufacturers have emerged over the last three years owing to the easy availability of the production technology. Thus, there are now maputi production sites in Epworth at Overspill, Chiremba, Solani and Reuben shopping sites. These maputi producers sell their products at a wholesale price of US$0.55 for a pack with 20 packets. Each packet then retails at US$0.10. The value chain for maputi is predominantly local and short, hence the price stability that is beneficial to the food system of the area. Moreover, the production of maputi is almost exclusively occurring in the informal sector, underlying the importance of the informal food sector to the overall food system of Epworth.

The last food tracked in the reverse value chain is offal. The association between poverty and consumption of offal has been documented (Mennell 1996). However, offal provides protein that is essential for food security and should therefore not be viewed negatively. In Epworth, offal meat was available from both the formal and informal food sector. There were, however, few butcheries in Epworth and the trade of offal was mainly by street traders as well as mobile vendors who would move around with buckets. The predominance of street traders was because they could sell small quantities for as low as US$0.50 while in formal butcheries the minimum purchase allowable was US$1.00. While offal meat is premised to be cheap and therefore affordable, the reverse value chain analysis in Epworth revealed that offal was generally beyond the reach of many residents, costing between US$4.00and US$5.50 per kg in comparison to beef that was trading at between US$5.50 and US$7 per kg. Under Epworth's stressed economic conditions, the food system was unable to supply residents with affordable foods necessary for a balanced diet.

Retail outlet location

Even in a food system that is dominated by an informal food sector, retail location is important as it usually determines the ease or difficulty with which consumers access food (Wilson *et al.* 2004), as well as consumers' purchasing patterns (Jago *et al.* 2007) and food choices (Larson and Story 2009). The absence of food retail outlets in particular areas may lead to the creation of an impoverished food environment where it is difficult to access nutritious and healthy foods (Bell *et al.* 2013). It is therefore important to understand retail outlet location in Epworth as this has a direct impact on the food security of the area's residents. The greatest proportion of the food retail outlets in Epworth were located in approved market areas in shopping precincts (38%). These approved markets do not sufficiently meet the demand for retail trading space however, and as such some retail outlets locate on street edges (18%) where they are able to capture customers from both vehicular and pedestrian traffic. In addition, some retail outlets were located within various shopping precincts (11%) in six of Epworth's seven wards. The exception was Ward Seven, which was entirely informal and had no designated shopping precinct or approved market areas.

The inadequacy of the regulated markets at shopping precincts in Epworth also means that a number of retail outlets were operating from illegal markets (11%) and homes (8%). This proliferation of illegal markets and the retail outlets in Epworth is not surprising, given that close to 70% of the residents live in largely unplanned and unserviced areas (Chenga 2010). Likewise, the operation of retail outlets from homes is common practice in Zimbabwe as it brings convenience to consumers who do not have to travel far to access goods (Chirisa 2009). Few retailers in Zimbabwe were operating from malls (0.5%). The only mall in Epworth, located along Chiremba road, is located away from most of the residential areas, requiring consumers to spend money on taxis, raising the cost of accessing food and thus lowering the likelihood of the consumer's patronage at them. For this reason, this mall was barely functional. Thus, over 60% of Epworth's food retailing was occurring outside approved market areas, with 89% of the retail outlets operating without a license, hence cementing the centrality of the informal food sector to Epworth's food system.

There are various reasons for the choice of location for retail outlets, including the maximization of the catchment area (Jaravaza and Chitando 2013). The findings of the Consuming Urban Poverty (CUP) research confirmed this, as the greatest proportion of food retail outlets in Epworth (36%) reported selecting locations so as to take advantage of opportunities offered by a location. Research also identified that about a quarter of the retail outlets (25%) were located at places that were deemed affordable, which validates Wood and Brown's (2006) assertion that location costs are a central consideration in food retail location. When considering more general economic issues, 15% of the outlets reported selecting a particular location as a consequence of general economic considerations, particular to their failed local economy, in which "dollarization" has meant that access to the US dollar is a key factor determining the viability of food retail locations, so reducing all operating costs is an essential survival strategy. The considerations of food retail entrepreneurs (of opportunities, economic costs and affordability) discussed in this section had a bearing on the spatial distribution of retail outlets, which will be discussed in the next section.

Typology and spatial distribution of retail outlets

Retail food outlets were classified in terms of broadly defined trader typologies and mapped. At a city level this mapping can show which outlets are found where, which foods they sell, and how they fit in the total food environment. The research results on the typology of the retail food outlets in Epworth indicate that street vendors (17%) dominate in the area, followed by grocery stores (14%) and tuck-shops (13%). The dominance of street vendors derives from the state of the economy in Epworth which, besides the few shopping precincts discussed earlier, has little trading infrastructure. Other significant typologies included butcheries (10%), grinding mills (10%), tables at intersections (7%) and house shops (6%). In addition to this, there were a (very) limited number (3%) of wholesalers identified in Epworth (see Figure 15.1). As Epworth lies within Harare's sphere of influence (20km away), much of this wholesale trade is dominated by large wholesalers from Harare where retailers can purchase goods at comparatively cheaper prices. Epworth's food system is thus intertwined with that of Harare's.

Two observations can be made from these findings, the first of which being that street vendors and tuck-shops dominate retail food trading. This is because street trading (and tuck-shops) requires a low capital investment to start the business (FAO 2003). There is also generally no bureaucratic process to follow for one to become a street trader and very little infrastructure required to commence trade. Additionally, the demise of Zimbabwe's economy pushed most people into street trading as a livelihood option (Ndiweni *et al.* 2014). Hence, the numerous street vendors and tuck-shops scattered across all the wards are extremely important to Epworth's food system. These street vendors sell a diversity of food items, ranging from fruits, vegetables, meat, *sadza* (a staple food made from maize-meal and served with relish), braai (barbeque) meat and cooked beans. The second observation is that there are no large-scale retail outlets in Epworth. This means that residents wanting to buy food from large supermarkets have to commute to Harare's residential area of Hatfield or to Harare City Centre, which in turn raises the cost of food acquisition, but further integrates Epworth into Harare's food system.

Food retail outlet ownership and income

The high unemployment rates that resulted from Zimbabwe's deteriorating economy have pushed both males and females into self-employment, particularly in the food retail sector (Ndiweni *et al.* 2014). The retail mapping results found that 64% of the food retail outlets were owned by females. This reflects statistics from the national Labour Force Survey (LFS) that indicate that there are more women than men active in the informal sector, particularly in the food service sector (ZIMSTAT 2014). The same LFS also showed that there were still more economically active men (30%) in paid employment than women (14%). Given the importance of the informal food retail sector in Epworth's food

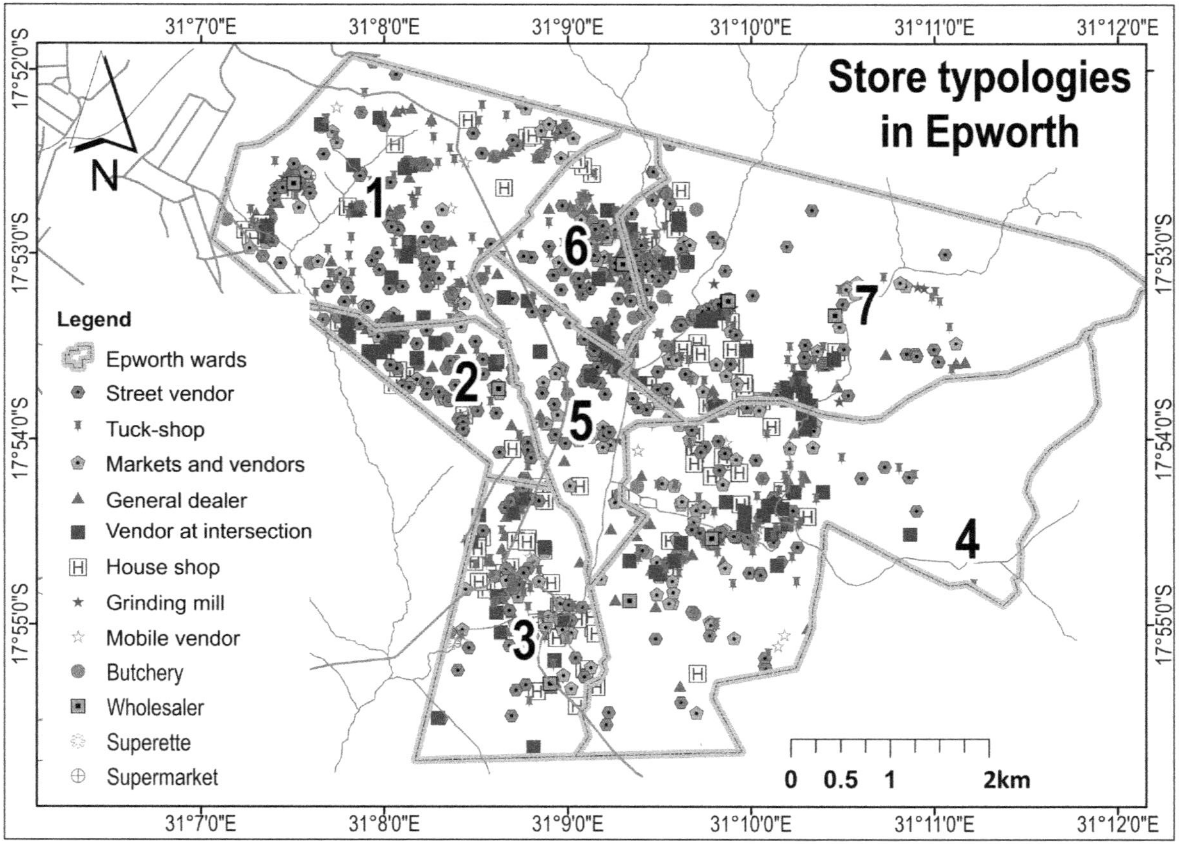

Figure 15.1 Spatial distribution of retail food outlets in Epworth

(Authors' own image, generated from Retail Mapping Survey – 14–21 Sept. 2016)

system, and women's numerical dominance in this sector, it is clear that women play a significant role in Epworth's food environment. Deteriorating economic conditions in the country have reduced employment opportunities and forced even the educated to eke out a living in the informal sector (Njaya 2015; Ndiweni *et al.* 2014). This statement was confirmed by the food retail surveys. While in the late 1990s statistics suggested that only 1.7% of traders had a full ordinary-level education (Mupedziswa and Gumbo 1998), the CUP survey found that retail outlet owners in Epworth were relatively educated, with 30% having completed senior secondary education. This change offers key insights into the changing nature of the Zimbabwean economy, and the difficulty of securing a formal job even with a good school education. The results also indicate that Epworth's food retail outlets are primarily under single ownership (62%) with the majority (81%) of food retail outlets being managed by their owners. This dominance of owner-operated food retail outlets could be explained by the fact that the business is both their source of employment and income (Bukaliya and Aleck 2012). The food retail system in Epworth could be characterized as being made up of a multiplicity of small players, the majority of whom operate within the informal food sector.

The informality that pervades Epworth's food system also extends to survival on income derived from businesses within this informal food sector. In the survey, 39% of operators indicated that their retail outlets were contributing more than 75% of their monthly household income, and 14% of the food retailers were deriving 51–75% of their monthly income from the retail outlets. Food retail is thus clearly important to the economy of Epworth. It is, however, also clear that the income derived from the retail outlets is not always sufficient to cover the needs of retail owners and their families. This is confirmed in the 86% of the retail outlet owners who indicated that the income was insufficient.

Food sold by retail outlets

In order to understand what was being sold by retail outlets in Epworth, the study classified the products into specific food retail categories. The most sold products are shown in Figure 15.2. The majority of these foods (sweets, sugared drinks, sweets, tea and coffee) are detrimental to health and high consumption may ultimately lead to lifestyle diseases such as diabetes and hypertension, as well as obesity, being overweight and poor oral health (WHO 2016).

The study findings are consistent with assertions made by Rimpeekool *et al.* (2017) that the diets of modernizing traditional societies are rapidly changing from low-fat cereal-based agrarian foods to industrial processed foods high in sodium and sugar. Thus, the consumption of such non-nutritive foods in Epworth is indicative of a society undergoing a "nutrition transition" which, however, has the potential to create significant health problems, especially non-communicable diseases, among the poor population that is resident in this area. The fact that the greater proportion of the trade in these foods occurs within the informal food sector is also problematic, given government and local

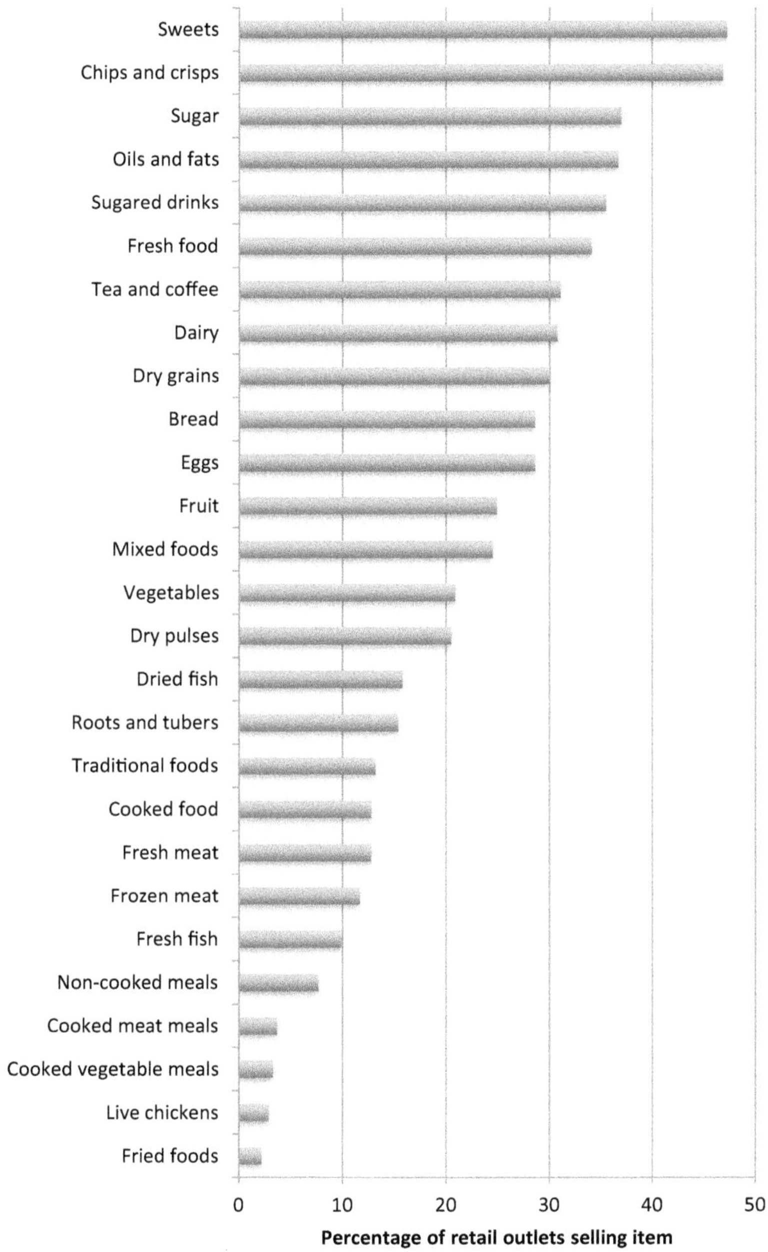

Figure 15.2 Food sold by retailers in Epworth

(Epworth Food Retailer Survey 2017)

authority's inability to police the sector in terms of health standards. This is even more imperative in Epworth, given that close to 40% of the retail outlets were operating without any form of license. Thus, while the informal food sector dominates the food system of Epworth, the risks posed by foods traded in this sector to human health should not be ignored. Rather the risks should be identified and flagged so that corrective and preemptive measures are put in place in order for this important component of Epworth's food system to operate safely.

Conclusion

The chapter sought to address two questions, the first on the major food sources in Epworth and how changes along the value chain enabled or impeded access, and the second on how the nature of retailing in the area was shaping the food system. Study findings revealed that Epworth's food system is entwined with that of Harare, while at the same time linked to regional and international markets through imports. Because of the absence of major shopping sites in the area, the informal food sector was generally dominating and therefore making crucial contributions to the food system of the area. While a wide variety of foods are sold in Epworth, non-nutritive foods such as sugared drinks and sweets were found to dominate the sales profiles of the retailers, raising questions about the health impacts that are likely to result, especially the burden of non-communicable diseases. Overall, Epworth's food system is dominated by small, largely informal players who provide a measure of service and food system access that is largely determined by factors outside their immediate control. This review of the food system offers useful insights into the food system of Epworth and also the broader economic and developmental challenges that the residents face.

References

Battersby, J. and Crush, J. (2014). Africa's urban food deserts. *Urban Forum*, 25(2), pp. 143–151.

Bell, J., Mora, G., Hagan, E., Rubin, V. and Karpyn, A. (2013). *Access to healthy foods and why it matters: A review of research.* The Food Trust. Published online: PolicyLink. Available at: http://thefoodtrust.org/uploads/media_items/access-to-healthy-food.original.pdf [Accessed 10 Nov. 2017].

Bukaliya, R. and Aleck, H. (2012). Challenges affecting informal business funding in Zimbabwe: Implications for the Zimbabwe open university. *The African Symposium*, 12(1). Available at: https://projects.ncsu.edu/aern/TAS12.1/TAS12.1Bukaliya.pdf [Accessed 14 Aug. 2017].

Chauvin, N.D., Mulangu, F. and Porto, G. (2012). Food Production and consumption trends in sub-Saharan Africa: Prospects for the transformation of the agricultural sector. *Working Paper 2012–011.* United Nations Development Programme. Available at: www.undp.org/content/dam/rba/docs/Working%20Papers/Food%20Production%20and%20Consumption.pdf [Accessed 22 Aug. 2017].

Chenga, N. (2010). Murambatsvina II Hovers over Epworth. *The Financial Gazette.* Available at: www.financialgazette.co.zw/national-report/3011-murambatsvina-ii-hoversover-epworth.html [Accessed 1 May 2018].

Chirisa, I. (2009). The geography of informal sector operations (ISOs): A perspective on urban Zimbabwe. *Journal of Geography & Regional Planning*, 12(4), pp. 66–79.

Crush, J. and Frayne, B. (2011). Supermarket expansion and the informal food economy in Southern African Cities: Implications for urban food security. *Journal of Southern African Studies*, 37(4), pp. 781–807.

FAO, (2017a). *Regional overview of food security and nutrition in Africa 2016: The challenges of building resilience to shocks and stresses*. Available at: www.fao.org/3/a-i6813e.pdf [Accessed 27 Feb. 2018].

FAO, (2017b). *Maize production forecast at bumper level in 2017, GIEWS*. Available at: www.fao.org/giews/countrybrief/country.jsp?code=ZWE [Accessed 21 Aug. 2017].

FAO, (2003). The informal food sector: Municipal support policies for operators. *"Food in Cities" Collection, No. 4*. Rome: Food and Agriculture Organization of the United Nations.

Gitz, V. (2016). Territorial food value chains for sustainable food systems: Initiatives. *French National food programme in sustainable value chains for sustainable food systems, A Workshop of the FAO/UNEP programme on Sustainable Food Systems*. Available at: www.fao.org/3/a-i6511e.pdf [Accessed 23 Jan. 2018].

Jago, R., Baranowski, T., Baranowski, J., Cullen, K. and Thompson, D. (2007). Distance to food stores and adolescent male fruit and vegetable consumption: Mediation effects. *International Journal of Behavioral Nutrition and Physical Activity*, 4(35), pp. 1–10.

Jaravaza, D.C. and Chitando, P. (2013). The role of store location in influencing customers' store choice. *Journal of Emerging Trends in Economics and Management Sciences*, 4(3), pp. 302–307.

Larson, N. and Story, M. (2009). A review of environmental influences on food choices. *Annals of Behavioral Medicine*, 38(1), pp. 56–73.

Mennell, S. (1996). *All manners of food: Eating and taste in England and France from the Middle Ages to Present*. New York: Blackwell.

Mupedziswa, R. and Gumbo, P. (1998). *Structural adjustment and women informal traders in Harare, Zimbabwe*. Uppsala: Nordiska Afrikainstitutet, No 106.

Muramba, F. (forthcoming). *Informal food processing and production: Practices of maputi producers in Harare's changing urban food system*. Unpublished Masters thesis, Department of Sociology, University of Cape Town.

Ndiweni, N., Ndiweni, J., Mashonganyika, S., Ncube, N. and Dube, N. (2014). Informal economy as a source of livelihood in Zimbabwean Urban Areas: The case of Bulawayo Metropolitan Province. *International Journal of Scientific and Research Publications*, 4(1), pp. 1–8.

Njaya, T. (2015). Informal sector, Panacea to the high unemployment in Zimbabwe? Case of informal sector enterprises of Harare Metropolitan. *International Journal of Research in Humanities and Social Studies*, 2(2), pp. 97–106.

Rimpeekool, W. Kirkk, M., Yiengprugsawan, V., Banewll, C., Seubsman, S. and Sleigh, A. (2017). Nutrition label experience and consumption of transitional foods among a nationwide cohort of 42, 750 Thai adults. *British Food Journal*, 119(2), pp. 425–439.

Swift, J. and Hamilton. K. (2001). Household food and livelihood security. In: S Devereux and S. Maxwell, eds., *Food Security in Sub-Saharan Africa*. London: ITDG Publishing.

Tacoli, C. (2016). *Informal food systems and food security in rural and urban East Africa*. IIED Briefing, February 2016. Available at: http://pubs.iied.org/pdfs/17336IIED.pdf [Accessed 22 Jan. 2018].

Tefft, J., Janasova, M., Adjao, R. and Morgan, A. (2017). Food systems for an urbanising world. *World Bank and Food and Agriculture Organization*. Available at: http://documents.worldbank.org/curated/en/454961511210702794/pdf/Food-Systems-for-an-Urbanising-World.pdf [Accessed 24 Jan. 2018].

Wilson, L.C., Alexander, A. and Lumbers, M. (2004). Food access and dietary variety among older people. *International Journal of Retail & Distribution Management*, 32(2), pp. 109–122.
WHO, (2016). *Reducing the consumption of sugar-sweetened beverages and their negative health impact in Estonia*. Copenhagen: WHO Europe Regional Office.
Wood, S. and Brown, S. (2006). *Convenience store location and forecasting—a practical research agenda*. Guildford: University of Surrey School of Management.
ZIMSTAT, (2014). *Labour force and survey, 2014*. Harare: ZIMSTA.

Part IV

The state of urban food poverty and its connections to the food system

16 Food poverty in Kisumu, Kenya

George Godwin Wagah, Nelson Obange, and Harun Okello Ogindo

Introduction

Poverty is a critical challenge for Kisumu. There have been many attempts made by government and development agencies to alleviate poverty in the city. This chapter argues that in order for poverty alleviation to be effective, it needs to be informed by an understanding of the multidimensional nature of poverty and that policies need to be informed by evidence. The experience of food security is argued here to be one means by which this multidimensionality can be examined.

Within official government articulations on food security, two dominant narratives are pervasive. The first is the presumed importance of urban agriculture as a food poverty relief strategy. The second is of a city whose population is increasing due to rural migrants to the city, who are eroding productive peri-urban land that could be used for production (see, for example, KNBS 2008). This chapter uses findings from the Consuming Urban Poverty (CUP) fieldwork to interrogate these assumptions.

While not representative of all the cities in the region, the earlier work of the African Food Security Urban Network, which investigated the food security status of selected, generally poor, communities in 11 Southern African cities, highlighted high levels of food poverty, particularly in poor neighbourhoods (Frayne *et al.* 2010). Through its discussion of the CUP findings, the chapter shows that food poverty is driven by more than just income poverty, supporting Mitlin and Satterthwaite's (2013) work on the multidimensionality of poverty and the notion that particular urban characteristics influence the scale and depth of poverty. The findings further illustrate that although food insecurity is most prevalent in poor areas of the city, it is a city-wide problem. The chapter then presents findings on the food source practices of households in Kisumu, highlighting the importance of market sources. The chapter concludes by returning to the dominant assumptions about potential solutions to food insecurity.

Information on the Kisumu context (its historical and economic development, its size and rate of growth, its socio-economic characteristics and spatial development) are contained in the Introduction to the book.

Poverty–food security nexus

Traditionally, poverty is defined in either relative or absolute terms. "Absolute poverty" measures poverty in relation to the amount of income necessary to meet the basic needs of a household, such as food, clothing, and shelter. The concept of absolute poverty, however, fails to recognize that individuals have important social and cultural needs. These limitations of the "absolute" framing led to the development of the concept of relative poverty. "Relative poverty" defines poverty in relation to the economic status of other members of the society: people are poor if they fall below prevailing standards of living in a given societal context (Foster 1998). An important criticism of both absolute and relative concepts of poverty is that they are largely limited to economic units of measurements, i.e. concerned only with binary measurements of income and consumption (UNESCO 2016). Furthermore, the traditional conceptualizations of poverty considered only monetary deprivation. Multidimensional Poverty Index (MPI) measures focus on a range of deprivation factors, such as poor health, lack of education, inadequate living standards, lack of income, disempowerment, poor quality of work, and threat of violence (Alkire 2010). These two more recent conceptualizations of poverty suggest that there is an urgent need to examine income poverty and urban food insecurity, but also to enquire into food poverty as more than just the outcome of constrained income.

In developing countries, particularly within sub-Saharan Africa, urban food poverty is argued to be driven by a variety of factors including rapid urbanisation, climate change, natural disasters, and inappropriate urban food system responses to global food system changes (Tendall *et al.* 2015). Paying attention to the nature and scale of food poverty in urban areas is not a traditional food poverty focus. When compared with rural areas general living standards tend to be better in cities; however, recent evidence suggests that urban poverty is on the rise in developing countries (Mitlin and Satterthwaite 2013). Multiple factors drive food poverty. Such factors include access to water and sanitation, spatial considerations, household structure, community networks and agency, and even factors such as a household's access to refrigeration (Battersby 2017). Urban centres in Kenya are no exception, and thus the challenges are not necessarily just hunger events and cycles that impact urban households for a short period of time, but also hunger events and cycles that appear to be more entrenched.

Research method

The 2009 Kenyan National Survey (KNS) data (KNBS/SID 2013) estimated the population of Kisumu to be 500 000, or 100 000 households. Allowing for a confidence interval of 10% at a 95% level of probability, the sample size required is 383 households (using the sample size table derived from the formula by Krejcie and Morgan 1970). To enable cross-tabulation between variables, the targeted sample size was 800. Ultimately 841 households were interviewed. This sample size provided a confidence interval of 7% at a 95% level of probability.

Kisumu can be divided into three discrete geographical zones: Western Kisumu (Zone A), the area west of the main A1 Road and north of the east/west B1 road; Eastern Kisumu (Zone B), located north of the B1 road and east of the A1 road; and the entire area south of the B1 road referred to as Southern Kisumu (Zone C). Colonial planning and geography of the city's growth mean that each zone contains a mix of high- and middle-income neighbourhoods, poor neighbourhoods, and peri-urban neighbourhoods. Neighbourhoods in each of these discrete zones were then purposefully selected, informed largely by ease of access (working relationships with local gatekeepers or chiefs, security, etc.) and distribution within the neighbourhood types (high-, middle-, and low-income neighbourhoods and peri-urban neighbourhoods). Two neighbourhoods of Kisumu were excluded from the sample. These were the Kaloleni neighbourhood with 3658 households and the Northern neighbourhood with 2107 households. Therefore, the sample represents the population of Kisumu excluding these two areas. The proportional nature of the sample means that it is self-weighting, so no weights needed to be calculated for the statistical analysis of the results.

The sample of households interviewed is proportional to the populations of the different geographical zones and different kinds of neighbourhood types (Table 16.1). In this respect, the sample estimates are representative of the different population sizes of these areas. The limitation of the sample design is that the three neighbourhoods selected within each neighbourhood type were not selected proportional to the size of their populations. So, to the extent that these neighbourhoods are not the same size, the sample is not an equal probability sample. Kisumu has three main informal settlement neighbourhoods – one each in the south, east, and north-west – and all these were covered in the survey. More than 65% of Kisumu reside in these neighbourhoods.

The selected residential areas were further divided into sections. A systematic sampling was done, targeting every third house along a road in the formal settlements and a household in each third plot in the informal settlements. Whenever an interview in the third house/plot was not possible the next house/plot was tried until an interview was successful. The questionnaire was administered to the head of the household, or a responsible adult in the household. The quantitative survey was supported by 50 in-depth household interviews distributed across Kisumu.

The survey questionnaire applied the instruments referred to as the Food and Nutrition Technical Assistance (FANTA) tools (Coates 2013), coupled with questions aligned to the Lived Poverty Index (LPI). The FANTA tools are cross-cultural household food security measures (Crush and McCordic 2017), including the Household Food Insecurity Access Prevalence (HFIAP) measure and the Household Dietary Diversity Score (HDDS). A key motivation for the use of FANTA and LPI was to ensure that the four major dimensions of food poverty (availability, accessibility, utilization, and stability, and their links to livelihoods and the operation of the broader food system) were measured.

Table 16.1 Number of households sampled by region

Geographical zones (A, B, & C) and their neighbourhoods	*Selected high- and middle-incomeneigh-bourhoods*	*Sample size*	*Population size (households)*	*Selected low-incomeneigh-bourhoods*	*Sample size*	*Population size (households)*	*Selected peri-urban neighbourhoods*	*Sample size*	*Population size (households)*	*Total population (households)*	*Total sample size*
Western (A): Tom Mboya, Robert Ouko, Mountain View, Obunga, Nyawita, Bandani, Kogony, Kanyakwar	Tom Mboya/ Mountainview/ Robert Ouko	26	3091	Nyawita/ Obunga/ Bandani	140	16651	Kanyakwar/ Kogony	45	5351	25093	211
Eastern (B): Kenya Re, Migosi, Kaloleni, Manyatta, Chiga, Nyamasaria	Kenya Re/Migosi/ Lolwe	42	4994	Manyatta A & B	241	28690	Chiga/Kasule/ Nyalunya/ Nyamasaria (north of highway)	40	4756	38440	323
Southern (C):Milimani, CBD, Lower Railways, Nyalenda, Kasagam, Nyamasaria (south of highway)	Milimani/CBD/ Railways	35	4162	Nyalenda A & B	224	26634	Kasagam/ Kasule/ Nyamasaria (south of the highway)	48	5707	36503	307
Total by types		103	12247		605	71975		133	15814	100036	841
% of total sample		1225	1224		7194	7195		1581	1581		

Lived experiences of poverty and food poverty in Kisumu

The average household size in the Kisumu sample was 4.16, with 49% of households being nuclear households, 22% female centred, 21% extended, and 7% male centred. The household roster data revealed two striking findings. The first was the youthfulness of the sample population. Well over half of the sampled population (56%) were under the age of 20, and over three-quarters (78%) were 35 or under. Just 2% were over the age of 65. The second was migrant status. Almost two-thirds (63%) of the sample population were born in Kisumu, and a further 12% were born in another urban area in Kenya. Despite the narrative of rural to urban migrants driving population growth, just 25% were born in a rural area.

The surveyed urban neighbourhoods in Kisumu have a mix of formal and informal settlement patterns which derive incomes from both formal and informal employment. Many Kisumu households rely on income from wage work in various forms. Just under a third of households (31%) reported some income from formal wage, 30% reported income from casual work (formal or informal), and 20% reported income from informal wage work. Households reported net income from a range of informal business types, but none of these sources were cited by more than 12% of households. Remittances, government grants and other sources of income were negligible. For employment and related labour market activities, of those of working age and eligible for employment (removing housewives, pensioners, and those with disabilities), 40% identified as self-employed, 27% reported working full time, 15% reported working part time, 13% reported being unemployed and looking for work, and 4% reported being unemployed and not looking for work. Despite the critical role that the informal sector plays in livelihood generation (Skinner 2016), particularly in the face of high or irregular employment, the high prevalence of self and informal employment impacts on the income-generating potential of households.

The vulnerable state of employment and the general state of the economy is reflected in the sentiment held by the respondents when discussing the state of the economy. When asked to compare their current livelihood situation to previous years, 13% of households reported being better off, 35% reported the situation to be unchanged, and 50% reported their situation being worse or far worse than the previous year.

Households were asked about their household expenditures for the month preceding the date of the interview – a 30-day recall exercise. Responses to this question offered insights into the spending profiles of households. Almost all households (96%) reported spending on food. Other expenditure categories cited by more than half of households were fuel (64%), housing (63%), transportation (58%), education (57%), and telecommunications (53%). While fewer households reported expenses such as debt repayments, cash remittances to rural areas, insurance, and the mutuality costs associated with family donations and support, these additional expenditure categories impact on already constrained household budgets.

In general terms, when compared to previous years, households in Kisumu face reduced economic opportunities. Informal work dominates livelihood strategies but the income-generating potential of these activities remains constrained. There are some formal (and stable) income-generating opportunities but these are limited and, as detailed by respondents, generally low-wage opportunities (examples of formal stable income-generating opportunities include: service work in hotels, entry-level clerical work, etc.). The need for payment for certain essential services, debt repayments, and other expenses that extend beyond the immediate household (such as family support) place additional pressures on the household budgets. Given the limited number of respondents who were comfortable to report on income, the income-based poverty analysis is limited. However, it is the contention of this chapter that monetary poverty is a very narrow way of understanding poverty, and that poverty is multidimensional and needs to consider alternative, more diverse means of assessing poverty. The survey therefore made use of the LPI as a further tool to better understand the poverty. It is worth noting though that the LPI "measures a portion of the central core of the concept of poverty not captured by existing objective or subjective measures" (Mattes 2008: 161). The chapter also argues that food insecurity provides another lens through which to view poverty.

The Lived Poverty Index

Developed by Afrobarometer, the LPI is an experiential measure, based on questions about how frequently people go without basic necessities during the course of a year. The LPI is a multidimensional poverty measure assessing access to food, fuel for cooking that food, water, energy, medical treatment, and income. As LPI scores increase, lived poverty increases. The LPI of sampled households in Kisumu seems quite low, with an average LPI score for the sample of 0.90, where 0.0 indicates that households were never without any of the categories and 4.00 indicates that households were always without all the categories. This is slightly better than the Kenyan average for 2014/2015 of 0.93 (Mattes *et al.* 2016: 7).

Deprivation is not uniform across the LPI categories (Table 16.2), with insufficient food to eat affecting 53% of households at some point in the previous year. Given the importance of the market as a source of food, the figure of 65% of households lacking a cash income during the previous year (Table 16.2) offers potential insights into the relationship between income and other household expenses. In addition to the lack of cash income, 57% of households had lacked enough water for home use, 50% enough fuel to cook food, and 50% medicine or medical treatment. These multiple deprivations shape a household's ability to ensure food security and dietary diversity. The costs of energy to cook food curtails food choice, as does access to water to clean and cook foods safely.

Table 16.2 Reported LPI household deprivations (n819) (% of households)

If ever, have you or your household gone without. . . :	*Never*	*Just once or twice*	*Several times*	*Many times*	*Always*
A cash income?	35	36	19	7	2
Enough fuel to cook your food?	50	33	14	3	0
Medicine or medical treatment?	50	32	14	3	0
Enough clean water for home use?	43	38	15	3	1
Enough food to eat?	46	31	19	4	0

Extent, depth, and nature of food insecurity in Kisumu

Access to food is a key dimension of food poverty. This is particularly relevant in the urban context. A diverse and varied food system exists in Kisumu (see Chapter 13). Kisumu is in the midst of a food system transition, with a number of supermarkets operating adjacent to street vendors; a varied collection of outlets selling prepared food; city approved and managed food markets (including fish, fresh fruit, and vegetables); and a clustering of more informal market spaces. Production in the broader region has undergone a transformation, from local and subsistence production to intensified cash-crop production, particularly sugar, meaning that most food is imported from elsewhere in Kenya or from other countries in the region. Kisumu and the wider county is therefore a net importer of food. Food is generally accessed through the market in Kisumu, with two-thirds of household reporting buying more than 75% of the food they consumed.

Urban food poverty is complex in nature: it depends not only on the buying capacity of urban citizens, but also on the ability to transport, store, preserve, and cook the foods they can afford to buy. The CUP study analyzed food availability and variety over a period of time to assess the extent of food poverty. The rest of this section will describe key food poverty indicators embedded within the suite of FANTA measures.

Levels of food insecurity, or food poverty, in Kisumu are generally high, with particularly high rates being measured in the poorer settlement areas. According to the HFIAP indicator, 71% of surveyed households in Kisumu were found to be moderately or severely food insecure, thus deemed to be experiencing food poverty. This figure, however, masks some place-specific challenges. Areas defined as peri-urban in the survey, generally those areas with higher levels of informality (and with reduced access to services), experienced particularly high levels of food insecurity. Eighty-two percent of the sampled Western peri-urban areas households, 87% of the Eastern peri-urban households, and as high as 91% of the Southern peri-urban households were moderately or severely food insecure.

The extent of food poverty is further compounded by poor dietary diversity. The HDDS tool is one measure used to assess the diversity of food consumed. The

HDDS is a "recall question" test to identify which foods in the 12 categorized food groups were consumed by anyone in the household in the past 24 hours. A HDDS score of six or less is a proxy indicator of likely malnutrition. Over three-quarters (78%) of households consumed food from six or fewer assessed food groups, with an average HDDS score of 4.05 items. The implication of this is a strong likelihood of both childhood stunting (which is an indicator of well-being and social inequality) and potential later challenges associated with obesity as a result of a low nutrient diet. The most commonly consumed food types were cereals and vegetables, specifically based on the traditional diet of *ugali* and *sukuma wiki* (green vegetables). Given the historical importance of fish within the Kisumu diet, it was surprising to see that relatively low proportions of households were consuming fish (28%) (Figure 16.1). The survey was conducted during a period of maize meal shortages in Kisumu (the base for *ugali*), thus it is possible that the high price of this essential foodstuff meant that households were sacrificing dietary diversity (such as fish) in order to meet basic food needs. Household interview respondents described how a meal without *ugali* was "not a meal", further emphasizing its importance from a broader food security perspective.

The low levels of dietary diversity coupled with the high levels of food insecurity place a real strain on the long-term development prospects of Kisumu. Given the youthful demographic of the Kisumu sample, the challenges of stunting (a result of reduced nutrition in the first 1000 days) means that the economic potential of the population will remain constrained while the risk of further diet-related illnesses is significantly increased. The consequence is that there is a real risk that poverty will remain entrenched. These food utilization challenges are further compounded by questions of household food supply

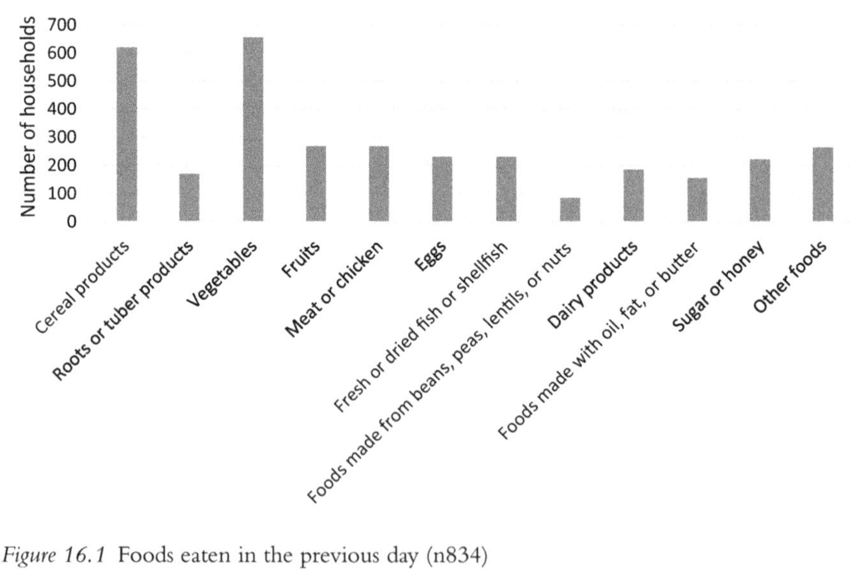

Figure 16.1 Foods eaten in the previous day (n834)

(Primary data. Author's own)

stability. While hunger seasons are not as extreme as in other cities (see Frayne *et al.* 2010) the Months of Adequate Household Food Provisioning (MAHFP) in Kisumu raises important questions.

The MAHFP statistics indicate relatively stable access to food. This indicator is based on the question, "Were there months, in the past 12 months, in which you did not have enough food to meet your family's needs?" It is therefore not about the quality of the food accessible, but about a total failure of access. The average MAHFP was 10.31, with over 80% of households having adequate access nine months or more. However, January was an outlier with 77.3% of households stating that this was a month where food access was severely constrained. August, September, and October were also challenging months but were not as extreme as January (Figure 16.2). The spike in January has been attributed largely to December festive season obligations, which include travel to certain familial areas and increased household spending, at times on costly credit, combined with the beginning of the school year.

These findings demonstrate a clear seasonality. In times of financial stress, poor households either chose to reduce certain foods, as a form of consumption smoothing, to ensure some basic foodstuffs in the house, or may simply not be able to afford certain foods. Households reported going without certain types of food because of the price of food (food being unaffordable) on a regular basis. Over a third of all households reported going without certain foods at least once a month, just less than 20% of households reported the same on a weekly basis, and 15% of households reported going without certain foods more than once a week. These figures reflect a highly unstable food

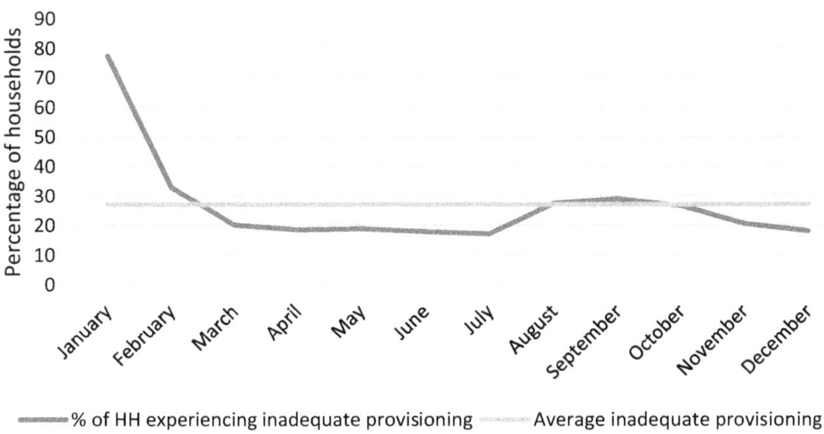

Figure 16.2 Percentage of households experiencing inadequate food provisioning (by month) (n831)

(Primary data. Author's own)

environment for households, an environment in which household food managers constantly have to discount certain foods in order to put food on the table.

When asked about the reasons for reduced access to food, respondents cited economic reasons linked to general household budget limitations as the key factors. The most significant reported causes of limited food access included reduced income of a household member (26% of households), loss of or reduced employment for household members (9%), and serious illness of a household member (8%). Reduced food access was also attributed to more socio-political factors, including insecurity and violence (13%) and the theft of money and food (9%). Households did not identify the absence of refrigeration, storage limitations, and increased costs of key services such as water and energy. One conclusion is that given the long-term lack of these assets, their potential impact on food security is not recognized.

One way in which households navigate the food system is to make use of multiple food access sources. Households are highly strategic in selecting food outlets used. A decision to frequent either so-called informal traders or new supermarkets in Kisumu is not just an impulse. These outlets are selected as a result of the value provided to the households, where value is not just about cost but factors such as storage in the household, neighbourhood dynamics (not appearing wealthier than others), local transport costs versus bulk savings, and many other factors. In relation to Chapter 13, which provides greater detail on the Kisumu food system, the next section considers key food access sources, offering some insights into the wider Kisumu food system.

Types of food retail outlets and household preferences in Kisumu

Varied food retail outlets characterize Kisumu. In the survey, households were asked about the frequency at which they accessed food from a range of sources, including a range of market-based sources, own production, and various forms of non-cash-based transfers, and the importance of these sources to their food consumption. The vast majority (88%) bought the majority of food the households consumed, demonstrating the importance of market sources. The only non-market source of food within ten most commonly used sources of food was food sent by relatives in rural areas, used by 7% of households in the last year (Table 16.3). Kiosks, market areas, and supermarkets reflect similar overall use, albeit with different use frequencies. Importantly the use frequency of the different outlets differs significantly with kiosks being frequented regularly (at least five times a week). Markets are commonly used as a weekly sourcing strategy, and supermarkets are used weekly or monthly as a source of food.

Such access strategies offer insights into the geography of food retail in Kisumu. The use frequency of the house shops and kiosks is informed by multiple factors, not just proximity to places of residence. Proximity is important given transport costs and other related costs; when the findings from the retail survey are considered, kiosks, market stalls, and even street vendors reported

Table 16.3 Frequency of accessing food from the ten most commonly used sources of food (multiple response options) (n840)

	At least five days a week	*At least once a week*	*At least once a month*	*At least once in six months*	*At least once a year*	*Total*
Kiosk/house shop in Kisumu	532	111	4	1	0	648
Supermarket	42	253	303	30	4	632
Market in Kisumu	145	366	59	1	0	571
Butcher in Kisumu	23	281	169	7	2	482
Grinding mill in Kisumu	6	161	185	7	0	359
Mini-market in Kisumu	109	107	7	0	0	223
Kiosk/tuck shop/vendor/ trader/hawker in Kisumu	101	30	2	2	0	135
Wholesaler in Kisumu	11	45	38	3	0	97
Restaurant in Kisumu	13	20	27	3	0	63
Food sent by relatives in rural areas in this country	0	1	17	35	7	60

offering credit. Over 64% of traders indicated that they usually or sometimes offered credit. Regular customers, known locals, and clients with good credit were the main recipients of credit. No supermarkets reported offering credit. These locally embedded retail options serve as a vital lifeline to those struggling with food poverty, enabling further food access.

Although market-based food sources dominated, there was some use of alternative mechanisms to source food, with a small proportion sourcing food from own production. Food was also provided by neighbours and other households in communities, meals shared with neighbours, and then a collection of remittance-type food receipts. While these alternative strategies were not widespread, they remain important mechanisms to ensure food security in times of income scarcity. Households also sought to smooth income to ensure access to food. Income-pooling strategies such as social savings groups are known to operate within Kisumu and are important in addressing food purchase challenges. In the study, 22% of households indicated that they participate in table banking (network savings schemes such as *stokvels* or *chilimba* networks – locally called *chama*) and savings as their strategy to navigate food deficits.

The policy discourse around food security in Kisumu centres on encouraging food production. However, just 1% of households reported sourcing food from growing in Kisumu in the past year, and 3% from livestock owned by the household. When asked why households did not produce their own food, the lack of land was the most commonly cited reason (59%). However, there were a number of other reasons cited by more than 30% of respondents, including: it is easier to buy than grow (46%), we have no time (38%), and people would steal what we grow (30%). Production currently plays a minor role in the household food sourcing strategies of very few households in Kisumu. The responses to the question of why households do not produce suggest that the reasons extend well beyond a simple lack of peri-urban land, as the policy discourse suggests.

Conclusion

By using food as a lens to interrogate the relationship that Kisumu residents have with the food system, this chapter has highlighted the interconnection between lived poverty, food insecurity, and a collection of lived realities in the Kisumu context. While an important contributing factor, poverty is not just about limited income. Households engage in a constant balancing act in order to meet the costs of everyday life, and food does not always take priority. Food is key costs but households are forced to allocate funds to a variety of sources and as a result often reduce diversity in diets or even consume less food. Food poverty was found to be higher in the poorer areas, particularly the peri-urban areas, but was not unique to these areas. Using the HFIAP score 71% of Kisumu are deemed to be moderately or severely food insecure. Despite its robust and arguably diverse urban food system, the extent of food poverty in Kisumu is high, as residents are unable to utilize these food system assets to ensure food security and a diverse diet.

These findings contradict some of the existing policy imaginings within Kisumu and Kenya more broadly. Urban agriculture cannot be presumed to be the desired and viable solution to food insecurity. The food source findings in conjunction with LPI findings suggest that the problem is more systemic and requires responses beyond providing access to land. Secondly, the concern that rural to urban migrants are occupying land that would otherwise be used for production is challenged by the fact that almost two-thirds of the sample population were in fact born in Kisumu. Although 40% of household heads were born in rural areas, only 25% of the sampled population were. This suggests that population growth in Kisumu is now more a result of natural population growth. This demands a reframing of policy discourse. While rural-to-urban migration as a consequence of rural under-development is generally argued to be a driver of urban poverty, the CUP findings did not support this. Challenging these framings is essential for policy and wider governance responses. The long-term developmental and health-related consequences of the observed food poverty places the community of Kisumu at risk of continued food insecurity and long-term under-development, well into the future.

References

Alkire, S. (2010). *Acute multidimensional poverty: A new index for developing countries*. Oxford: Oxford Poverty and Human Development Initiative.

Battersby, J. (2017). Cities, planning and urban food poverty in Africa. In: V. Watson, G. Bhan, and S. Srinvas, eds., *Companion to planning in the global south*. London: Routledge, pp. 204–214.

Coates, J. (2013). Build it back better: Deconstructing food security for improved measurement and action. *Global Food Security*, 2(3), pp. 188–194.

Crush, J. and McCordic, C. (2017). The hungry cities food purchases matrix: Household food sourcing and food system interaction. *Urban Forum*, 28(4), pp. 421–433.

Foster, J.E. (1998). Absolute versus relative poverty. *The American Economic Review*, 88(2), pp. 335–341.

Frayne, B., Pendleton, W., Crush, J., Acquah, B., Battersby-Lennard, J., Bras, E., . . . and Leduka, C. (2010). *The state of urban food insecurity in southern Africa*. Urban food security series, 2. Cape Town: AFSUN.

Kenyan National Bureau of Statistics and Society for International Development (KNBS/SID), (2013). *Exploring Kenya's inequality: Pulling apart or pooling together*? Nairobi: KNBS and SID.

Kenyan National Bureau of Statistics (KNBS), (2008). *Constituency report on well-being in Kenya: Report based on the Kenya integrated household budget survey-2005/06*. Nairobi: Government of Kenya.

Krejcie, R.V. and Morgan, D.W. (1970). Determining sample size for research activities. *Educational and psychological measurement*, 30(3), pp. 607–610.

Mattes, R. (2008). The material and political bases of lived poverty in Africa: Insights from the Afrobarometer. In: V. Møller, D. Huschka, and A.C. Michalos, eds., *Barometers of quality of life around the globe. Social Indicators Research Series*, Vol 33. Dordrecht: Springer, pp. 161–185.

Mattes, R., Dulani, D. and Gyimah-Boadi, E. (2016). *Africa's growth dividend? Lived poverty drops across much of the continent*. Afrobarometer Policy Paper No. 29, January 2016. Available at: http://afrobarometer.org./sites/default/files/publications/Policy%20papers/ab_r6_policypaperno29_lived_poverty_declines_in_africa_eng.pdf [Accessed 1 May 2018].

Mitlin, D. and Satterthwaite, D. (2013). *Urban poverty in the global south: Scale and nature*. Abingdon: Routledge.

Skinner, C. (2016). Informal food retail in Africa: A review of evidence. *Consuming Urban Poverty Project, Working Paper 2*. Cape Town: African Centre for Cities.

Tendall, D., Joerin, J., Kopainsky, B., Edwards, P., Shreck, A., Le, Q.B., . . . and Six, J. (2015). Food system resilience: Defining the concept. *Global Food Security*, 6, pp. 17–23.

UNESCO, (2016). *International migration*. Retrieved March 12, 2017. Available at: www.unesco.org/new/en/social-and-human-sciences/themes/international-migration [Accessed 15 Mar. 2017].

17 Food poverty in Kitwe, Zambia

Issahaka Fuseini and Owen Sichone

Introduction

Increasing urban poverty is one of the visible features of current urbanisation dynamics in sub-Saharan Africa and urban food poverty is a manifestation of the multifaceted, multi-scalar nature of this evolving urban poverty (IFPRI 2017). The Consuming Urban Poverty (CUP) project therefore sought to scrutinize food poverty as an indicator of the workings of persistent, pervasive, and systemic urban poverty. This chapter presents the findings of household food poverty research in Kitwe, Zambia.

Background

The economic history of Kitwe is explained in the Introduction to the book. Significant changes in the copper mining industry over the years have affected rates of urbanisation as well as levels of employment, poverty, and food security. The increased exposure of residents in the city to food poverty also happens within the context of high national food insecurity levels, with 45.9% of Zambia's population reported to suffer undernourishment. This statistic places Zambia as the fifth hungriest nation in the world (IFPRI/Welthungerhilfe/ConcernWorldwide 2017).

The CUP project's food poverty survey in Kitwe was designed as a pro-poor survey, with purposive selection of two poor neighbourhoods. The results presented in this chapter are therefore not generalisable for the entire city. Nevertheless, the food poverty situation in the two poor neighbourhoods resonates with the food poverty conditions of other urban poor households in the city as shown in an earlier study by Mwitwa and Ng'andwe (2010). As presented in Chapter 3, the food security status of urban residents hinges on complex factors and processes that determine availability, accessibility, stability, and utilisation of food. Put differently, the ability of the urban households to access food regularly at the right quantities and quality transcends mere availability of food to encompass other structural systems such as the broader food system, other urban systems, socio-cultural variables, or non-food factors, as well as spatial characteristics of the urban settlements.

This research adopts the 1996 Food and Agriculture Organization's (FAO) definition of food security as "A situation in which all people, at all times, have physical, social and economic access to sufficient, safe and nutritious food which meets their dietary needs and food preferences for an active and healthy life" (FAO 1996). Food poverty is considered to be the absence of food security. The chapter illustrates household food poverty in the study neighbourhoods and argues that food poverty among the urban poor in Kitwe is due to general income poverty underpinned by structural macroeconomic factors beyond the control of the household, though some household-scale factors such as the size and structure of households contribute to the magnitude of the phenomenon. It concludes that policy and programmatic responses ought to engage with the food system and livelihood activities of the urban poor in order to build their adaptive capacity to poverty and food insecurity.

Research method

The findings discussed within this chapter are based on the household questionnaire survey conducted in Kitwe in March 2017, which measured the extent of household food poverty and its various aspects. A limited number of in-depth qualitative interviews were also carried out to help inform the questions in the household survey and to gain some insights into how households navigate food poverty, but the chapter does not draw on this directly. The household food security study in Kitwe was designed as a pro-poor survey. The population to be surveyed was defined by choosing two poor neighbourhoods in the city: hence the questionnaire survey aimed to represent the numbers in these two neighbourhoods – Ipusukilo and Wusakile. The survey results therefore cannot be generalised to the whole of Kitwe.

Given the lack of up-to-date official data at the suburban scale, the research team drew on "grey" material from the municipality to develop its sample frame. The municipality made available household count data that were generated in the two neighbourhoods as part of a settlement upgrading project in 2016. In total, the household count for Ipusukilo was 4717 and for Wusakile 4663, a combined total of 9380 households. Using the combined total number of households and the target sample size of 800 households for Kitwe, the team generated representative samples that were proportional in size to the number of households in each of the two communities. Thus, samples of 402 and 398, respectively, were determined for Ipusukilo and Wusakile. An additional 51 and 28 households, respectively, were surveyed in anticipation of incomplete responses. This brought the actual sample sizes to 453 in Ipusukilo and 426 in Wusakile, and the total sample to 879. These figures provide 95% confidence intervals of 9% and 10%, respectively, assuming the maximum variance in the population parameters (Czaja and Blair 1996).

The survey used a systematic sample with a random start, determined by a random number generator app on the tablets used to collected the data. Following completion of a questionnaire, the enumerator moved on three houses. In a case of no response, the enumerator went to the house directly opposite the house with no response. If no-one in this house was available, the random generator was used to identify the next house.

The core focus of the questionnaire was the Food and Nutrition Technical Assistance (FANTA) project's food security indicators and the Lived Poverty Index (LPI). The indicators include the Household Food Insecurity Access Prevalence Indicator (HFIAP), the Household Dietary Diversity Scale (HDDS), and the Months of Adequate Household Food Provisioning Indicator (MAHFP). Further details on these indices can be found in Swindale and Bilinsky (2006), Coates *et al.* (2007), and Bilinsky and Swindale (2010).

Socio-economic characteristics of households

The following section provides an overview of the socio-economic characteristics of sampled households in order to provide a context for the interpretation of the food security findings.

Household structure, size, and age

Four household types were identified within the two neighbourhoods: nuclear households, extended households (a household which extends beyond the nuclear family), female-centred households (a household with a female head and no adult male present), and male-centred households. Almost half of the households (48%) were nuclear, with the next most common being extended (27%), followed by 22% female centred. Just 4% were male centred – a sharp contrast to the early demographic of Kitwe as a mining town dependent on single male migrants. As will be presented later in the chapter, household structure appears to be a determinant of poverty and food security.

The average household size in the study neighbourhoods was 5.54 people with a maximum household size of 13 members. This average size was slightly higher than the national average household size of 5.1 determined in Zambia's 2013–2014 Demographic and Health Survey (CSO-Zambia *et al.* 2014). At the neighbourhood level, Wusakile had an average household size of 5.71 compared to Ipusukilo's 5.38. Female-centred households were slightly larger on average than nuclear households (5.33 and 4.87, respectively), with the average size of an extended household being considerably larger at 7.07.

The age of the sampled population indicates a youthful demographic amongst poor Kitwe households. The average age of sampled household members was 22 years old, and over 60% of the sampled population was under 25. A youthful population tends to have high levels of dependency in terms of many young members not earning an income, yet they tend to have higher expenditure on items such as health and education.

Education, employment, income, and migration

Educational attainment of household members was very low. More than half of all household members interviewed never completed primary school (with 20% having no formal school at all) while about 29% completed primary school and/or had some high school education and only 13% had completed high school and/or had post-secondary qualifications. The low levels of education in the communities means that even if formal skilled job opportunities were available many of the residents would be unable to access these. The majority of working-age (15 years and over) people included in the sample identified as self-employed (38%). The remainder were full-time workers in the private sector (19%) or worked part time or seasonally (16%).

Obtaining information on household income in the survey was challenging due to a very high non-response rate (44%). The mean monthly income from all sources for the households that responded to the question on income was ZMW1384.33 (at the time of the survey in March 2017 one Kwacha was worth US$0.10), although there was a significant range with almost ZMW25 000 between the highest and lowest reported incomes. The mean household monthly expenditure was ZMW1085.46, which leaves a low "net" (surplus) income. This is not sufficient to allow for household savings, capital formation, or to perform other social obligations such as remittances. Only 3% of the sampled households indicated that they save from their monthly income, saving an average of ZMW710.16. Mean household expenditure on food and groceries was ZMW614.96, which represented 44% of mean household monthly income. Education (ZMW531.68) and housing (ZMW300.86) were other items that a significant proportion of the mean monthly household income was spent on. The competing demands on household income complicates the food and income poverty of urban households, especially against the backdrop of precarious household income. Over two-thirds of sampled households stated that their economic conditions were much worse (42%) or worse (27%) than the previous year.

Over 80% of all household members sampled were born in Kitwe. The larger part of the migrant population (11% of the sampled population) had migrated from other urban centres in Zambia, with just 7% identifying as having moved from rural areas within the country, and a negligible proportion coming from outside of Zambia. Migration is therefore not an important characteristic of the sampled population of Kitwe, and thus was not seen as an important factor when it came to interpreting the food security findings.

Extent of household food insecurity

This chapter argues that through focusing on both the lived experience of food insecurity and how households seek to maximise food security, it is possible to develop a better understanding of the nature and drivers of poverty. This section therefore presents findings from the food security survey, contrasting findings from the commonly used LPI with findings from FANTA's suite of indicators.

Lived Poverty Index

The LPI was developed by Afrobarometer as a measure to try and quantify the experiential nature of poverty by asking about access to food, water, medicine or medical treatment, fuel to cook, and cash incomes. The LPI is therefore an attempt to move beyond income-based measures of poverty, and is argued to offer "an important complement to official statistics on poverty and development" (Mattes *et al.* 2016: 2).

On the surface, the lived poverty of sampled households in Kitwe seems quite low, with an average LPI score for the sample of 1.34, where 0.0 indicates that households were never without any of the categories and 4.00 indicates that households were always without all the categories. This is a little higher than the Zambian average for 2014/2015 of 1.14 (Mattes *et al.* 2016: 7). The average LPI for female-centred households was found to be the highest of all the household types (1.51), which confirms earlier work on the poverty of female-headed households in Kitwe (Mwitwa and Ng'andwe 2010).

Deprivation is not uniform across the LPI categories, with insufficient food to eat affecting all but 4% of households at some point in the previous year, and all but 12% of households lacking a cash income during the previous year (Figure 17.1). The relatively high proportion of households lacking food compared to the other indicators is perhaps indicative of the fact that households may be able to source water, fuel, and medical services from the state or by non-monetary means, but access to food is dependent on access to money. Previous food security research has found that households treat food as one of their discretionary expenditures, and will temporarily sacrifice access to food in order to meet other household needs like rent and debt repayments (Smith and Abrahams 2016).

The composite LPI score seems to indicate relatively low levels of poverty with the sample population, but interrogation of the categories suggests that food insecurity levels are high. The LPI may therefore be a useful baseline indicator to gauge levels of poverty at a population level, but in order for it to have utility for practitioners and policy makers it is necessary to move beyond the composite LPI scores to the level of the constituent parts. It is at the level of the constituent parts that the complexities of the competing variables in the household budget can be detected and appreciated in ways that enable evidence-based targeting.

Food and Nutrition Technical Assistance project's indicators

The LPI findings suggest high levels of food poverty. It is the contention of the CUP project that food poverty is multidimensional and dynamic. However, many studies assess food insecurity on the basis of a single indicator. This chapter argues that by engaging a suite of indicators, a more nuanced understanding

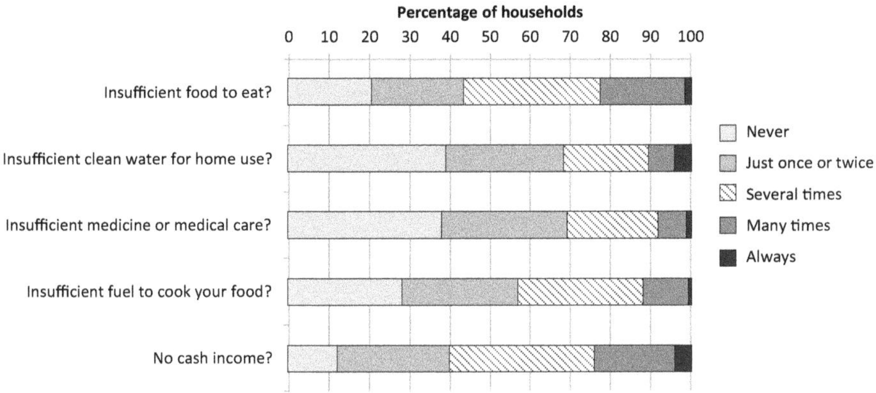

Figure 17.1 Percentage of households experiencing LPI constituent elements

(Author's own)

of the nature and dynamics of food poverty can be obtained. The following section therefore considers the findings of the Kitwe survey with reference to the three FANTA food insecurity indicators.

Household Food Insecurity Access Prevalence levels

The results of the study revealed high levels of household food poverty among the sampled population. All three FANTA indictors revealed widespread food poverty in the two study neighbourhoods. The HFIAP indicators (which is a computed indicator that categorises households as "food secure", "mildly food insecure", "moderately food insecure", and "severely food insecure") found that 79% of the households surveyed were severely food insecure. When the moderately food insecure category is added to the severely food insecure, the proportion of food insecure households comes up to over 90%. Only 6% of the study households were found to be food secure. A cross-tabulation of household size with household food security status suggests that households with higher average sizes were more food insecure than those with lower average sizes.

Food security in researched communities is principally related to household income and the ability to access food through purchase. Given that urban food security is driven largely by access rather than availability, the ability to purchase food by the households is closely linked to prevailing economic conditions that enable or constrain employment status and household income (Mwitwa and Ng'andwe 2010; Battersby 2012). As a result, the aforementioned downturn in the mining industry has had negative effects on employment, household incomes, and therefore their ability to acquire food (Fraser and Lungu 2007). In the household survey households were asked to enumerate the various shocks that constrain their ability to acquire food, and their responses confirmed urban residents' dependence on income and employment for their food needs. The most frequently identified shocks were: reduced income (31%), loss/reduced employment (26%), death of family member (14%), and sickness of family member (13%). The impacts of these individual shocks are not offset by even the recent national surplus in the production of maize (the main staple in the country), reaffirming that the availability of food does not necessarily solve the problem of food insecurity for poor households.

Household Dietary Diversity

The HDDS tool measures household food insecurity through scores based on dietary diversity, whereby a score of six or below is a proxy indicator for malnutrition. The severity of household food insecurity in the two communities is well reflected in the HDDS scale, as it showed that dietary diversity is very limited for most households in the study. In Kitwe, the mean HDDS was 3.25, and female-centred households displayed a slightly lower dietary diversity than male-centred ones (3.05 compared to 3.39). These figures indicate that diets are severely restricted, which has significant long-term health and development impacts. This reflects national circumstances, as Zambia has some of the poorest anthropometric indicators of food security in the world, with, for example, a national level of stunting in children under the age of 5 at 45% (UNICEF 2008). Mwitwa and Ng'andwe's study highlights this at the metropolitan scale, finding levels of stunting in Kitwe to be at 48%, which is marginally higher than the national average (Mwitwa and Ng'andwe 2010).

Table 17.1 provides indication of the kinds of food respondents consumed. The four varieties that most respondents reported consuming were cereals (mainly maize) (at 98%), vegetables (83%), fish (33%), and meat or chicken (24%), while the least consumed foodstuffs reported were fruits (4%) and dairy products (4%). These consumption patterns reflect a typical Zambian diet characterised by regular consumption of maize meal (*nshima*) with vegetables. While maize and vegetables seem indispensable, fish and meat are commonly added as a supplement for protein but these can be done away with if the household operates under a very tight budget (Mwitwa and Ng'andwe 2010). When respondents were asked which food types households lacked in the past six months, meat/chicken (78%), dairy products (51%), and fish (38%) were the most prevalent. These figures confirm Mwitwa and Ng'andwe's findings on food consumption patterns, as it indicates that in times of financial difficulty protein sources are excluded in order to maintain access to the staple foods of maize meal and vegetables. This practice was found to be used more frequently in the aftermath of the global financial crisis in the 2008/2009 period, as Mwitwa and Ng'andwe (2010) found that poor urban households in Kitwe

Table 17.1 Categories of foods eaten by surveyed households (%)

Food categories	*Yes (%)*	*No (%)*
Cereal products	98	2
Roots or tuber products	5	95
Vegetables	83	17
Fruits	4	96
Meat or chicken	24	76
Eggs	13	87
Fresh or dried fish or shellfish	33	67
Foods made from beans, peas, lentils, or nuts	15	84
Dairy products	4	96
Foods made with oil, fat, or butter	14	86
Sugar or honey	14	86
Other foods	17	83

resorted to reduced quantity and quality of their food, often depending primarily on maize meal.

The importance of maize in the diets of the poor is reflected in the national government discourses on food security and agriculture. National discussions on food security are centred on the "crop", such that crop availability is synonymous with food security and vice versa. National government therefore prioritises production and distribution of maize over all other crops (Mwitwa *et al.* 2016). This orientation has its roots in the colonial legacy of using scarcity and food to assert state control over its people and territory. For example, Duminy (2017) and Chapter 5 shows how the colonial authorities in British East and southern Africa employed a combination of agricultural and marketing policies to fight the scarcity of food which ultimately resulted in the single-minded promotion of maize. However, in all this, household income relative to prevailing food prices still plays an important role in determining household dietary diversity. Therefore, the patterns of household food consumption in the study communities are both the functions of political economy (which shapes the relative availability of different types of food) and affordability.

Months of Adequate Household Food Provisioning

While the HFIAP and HDDS scores indicate high levels of food insecurity and malnutrition, the MAHFP statistics indicate relatively stable access to food. This indicator is based on the question, "Were there months, in the past 12 months, in which you did not have enough food to meet your family's needs?" It is therefore not about the quality of the food accessible, but about a total failure of access. The average months of adequate provisioning was 8.95, with almost three-quarters of households having adequate

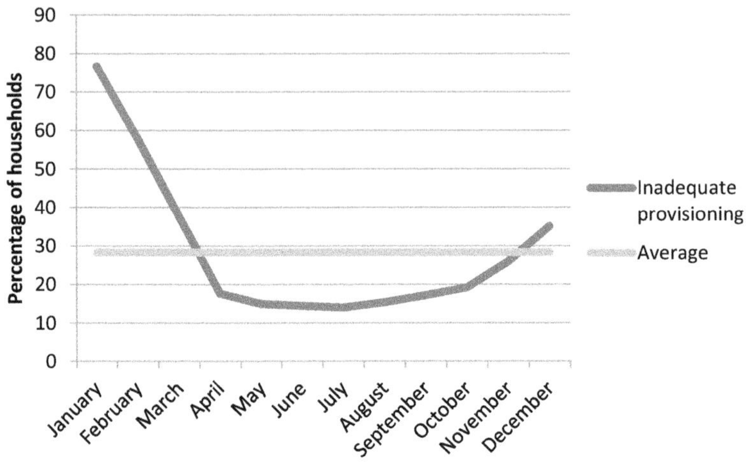

Figure 17.2 Percentage of households experiencing inadequate food provisioning (by month) (Author's own)

access nine months or more (74%). Female-centred households were again the most vulnerable, having an average of just 7.96 months of adequate provisioning.

Figure 17.2 indicates that there is a distinct seasonality to household access to food, with 77% of households reporting inadequate access to food in January, and less than 20% of households having inadequate access in the months from April to October. This suggests that adequate provisioning is related to the high dependence on maize, with the months after the maize harvest, when prices are lower, being the periods of least limited access. It is also likely that the high access failure in January is because of households deferring food expenditure to meet other household needs and associated debts that relate to year-end obligations.

How do households navigate food insecurity in Kitwe?

The data above indicate that food poverty in Kitwe is characterised by limited dietary diversity and distinct periods of covariate food insecurity during the year, which correspond with periods of limited accessibility of maize meal. It is important to understand how households navigate their food system in order to maximise their food security. Households accessed food from a variety of sources, but the most frequent sources of food were local neighbourhood markets, local vendors (ka table/vendor/hawker), and houseshops in the local neighbourhood, used by 70%, 39%, and 31% of households at least once a month, respectively. When supermarkets were used, they were used infrequently (just 6% used them weekly or more frequently). A pattern appears of daily and weekly purchases being made within the neighbourhood, and monthly purchases being made at supermarkets, markets outside the neighbourhood (e.g. Chisokone Market), and wholesalers within the neighbourhood.

More than half of surveyed households (53%) indicated that they got less than a quarter of their food from supermarkets, and 39% report obtaining less than 10% of the food they consumed from these retailers. Almost three-quarters of households (73%) sourced more than half of their food from their immediate neighbourhood. This indicates high dependence on local markets and vendors as key sources of food in Kitwe.

These purchasing patterns reflect the responsiveness of the local food retail sector to the lived needs of the urban poor, as described in Chapter 14. Small-scale local retailers are physically accessible, sell food in unit sizes affordable to the poor, have long opening hours, and at times offer food on credit. However, consumers will still make less frequent purchases of non-perishables from the supermarkets, larger markets, and wholesalers which offer relatively lower per unit prices but often sell in bulk (Battersby 2012). This allows poor urban households to plan and buy in bulk but infrequently (Skinner and Haysom 2016), which is why most of households that reported buying food from the supermarkets did so infrequently. Households therefore navigate their food system strategically to meet their food needs as best they can.

Non-market sources of food were extremely low. Just four out of 879 households were getting food from their own livestock, five from their own production outside of Kitwe, and five from their own production within Kitwe. Only 2% of sampled households were receiving food from relatives in the form of remittance, and only 2% were receiving food through sharing with neighbours, borrowing from neighbours, or getting food from neighbours by any other means. Only two out of the 879 participating households reported receiving food from work. This last point is particularly striking given that prior to the privatisation of mining, mine workers usually obtained food through the mines (Fraser and Lungu 2007). This source of food would have played a significant role in household food security, and its loss therefore may have driven many mining households into food insecurity.

Given the high dependence on market-based sources of food, it is important to consider not just the ways in which households diversify purchasing strategies to maximise food security, but also the ways in which they seek to maximise their income to achieve food security. The mass retrenchments across the mining sector have meant that there are few formal employment and income opportunities (Fraser and Lungu 2007). Just 26% of the respondents indicated getting their main household income from formal sector work during the past month, compared to 44% from informal sector and 17% from casual work. Mwitwa and Ng'andwe (2010) observed rapid growth in the informal sector in Kitwe following the 2008/2009 global financial crisis. The household survey included 4868 people in total, including 2044 who were still in school or too young for school. Of those of working age, by far the most common source of employment was "self-employed" (N=546), with just 108 identifying as mine workers. Household members therefore are highly dependent on self-employment as a means to support household needs in the absence of formal employment opportunities.

Conclusion

Households in Kitwe, as elsewhere in the global South (IFPRI 2017), are dependent on cash incomes to access food. Due to the precarious nature of urban livelihoods and associated incomes, global and national economic trends have precipitated and entrenched shocks within urban economies. For Kitwe, these shocks manifested in the form of structural changes in the mining sector, exposing many poor urban households to a high incidence of poverty and food insecurity. These structural and macroeconomic factors that mediate household food insecurity reaffirm the inadequacies of treating household food poverty at the household scale or with productionist lenses (Battersby 2012). Households in Kitwe have little or no control over the processes that shape their food insecurity situation. Rather, it is the combination of structural macroeconomic factors (Fraser and Lungu 2007) and colonial legacy (Duminy 2017 and Chapter 5) that have shaped the nature of food poverty among the city's poor households.

The food insecurity data presented in this chapter indicate household poverty manifesting as severe food insecurity characterised by extremely limited dietary diversity. While diets were extremely limited, the months of inadequate food provisioning were fewer than the HFIAP figures would usually suggest for a city in the global South. This is the result of national food policies that focus on ensuring production and access of maize. The agriculture and food policies of the Zambian government have provided relatively stable accessibility of maize as a staple food to urban households throughout the years. However, this privileging of maize production and distribution has led to the neglect of non-staple crop production and marketing support. This has led to relatively expensive domestic food production and supply, and dependence on costly imports. The outcome of these agriculture and food policies and programmes is that most households can access a basic diet of maize throughout the year (hence the high MAHFP), but are unable to afford a diverse diet. In contexts like Africa, where food insecurity is often conflated by policy makers with hunger, the constant supply of maize provides a veneer of food security, but masks significant nutritional deficiencies.

This chapter therefore argues that by using the suite of FANTA indicators, instead of a single food security indicator, it is possible to demonstrate how household food security and nutrition are determined not just by household income poverty and living conditions but also by wider agricultural policy decisions. Additionally, the LPI may be a useful baseline indicator to gauge levels of poverty at a population level; however, in order for it to be meaningful for practitioners and policy makers it is necessary to move beyond the composite LPI scores and examine individual deprivations.

The findings of this chapter indicate that households navigate food poverty through strategic engagement with the city's food system and livelihood diversification. This is framed by the wider structure of the food system and the urban system. Therefore, policies and programmatic responses that are meant to alleviate food poverty ought to engage both the food system and livelihood interventions so as to promote access to food and enhance household income.

References

Battersby, J. (2012). Beyond the food desert: Finding ways to speak about food security in South Africa. *Geografiska Annaler, Series B: Human Geography*, 94(2), pp. 141–159.

Bilinsky, P. and Swindale, A. (2010). *Months of adequate household food provisioning (MAHFP) for measurement of household food access: Indicator Guide (v. 4).* Washington, DC: FHI 360/ FANTA.

Coates, J., Swindale, A. and Bilinsky, P. (2007). *Household food insecurity access scale (HFIAS) for measurement of food access: Indicator guide (v.3).* Washington, DC: Food and Nutritional Technical assistance Programme and Academy for Educational Development.

CSO-Zambia, MOH-Zambia, and ICF International, (2014). *Zambia demographic and health survey 2013–14.* Rockville, MD: Central Statistical Office/Zambia, Ministry of Health/ Zambia, and ICF International.

Czaja, R. and Blair, J. (1996). Designing surveys: A guide to decisions and procedures. Thousand Oaks, CA: Pine Forge Press.

Duminy, J. (2017). Ecologizing regions; securing food: Governing scarcity, population and territory in British East and Southern Africa. *Territory, Politics, Governance.* doi:10.1080/21622671.2017.1306457

FAO, (1996). *World food summit plan of action Rome, 13–17 November.* Available at: www.fao.org/docrep/003/w3613e/w3613e00.HTM [Accessed 24 Jan. 2018].

Fraser, A. and Lungu, J. (2007). For whom the windfalls? Winners & losers in the epri-vatisation of Zambia's Copper Mines. Lusaka: Civil Society Trade Network of Zambia (CSTNZ) and Catholic Commission for Justice Development and Peace (CCJDP).

IFPRI, (2017). *2017 global food policy report.* Washington, DC: International Food Policy Research Institute.

IFPRI/Welthungerhilfe/ConcernWorldwide, (2017). *2016 Global Hunger Index: The inequalities of hunger.* Washington, DC/Dublin/Bonn: IFPRI.

Mattes, B.R., Dulani, B. and Gyimah-Boadi, E. (2016). *Africa's growth dividend? Lived poverty drops across much of the continent.* Afrobarometer Policy Paper, No. 29, January 2016. Available at: http://afrobarometer.org./sites/default/files/publications/Policy%20papers/ab_r6_policypaperno29_lived_poverty_declines_in_africa_eng.pdf [Accessed 1 May 2018].

Mwitwa, J. and Ng'andwe, P. (2010). *Effects of the global financial crisis on the food security of poor urban households: Case study Kitwe, Zambia.* Zambia: RUAF, Copperbelt University.

Mwitwa, J., Sibajene, M., Chipoya, G.C. and Namiluko, Y. (2016). *City region food system situational analysis, Kitwe, Zambia.* Rome: FAO—Food for Cities Programme.

Skinner, C. and Haysom, G. (2016). The informal sector's role in food security: A missing link in policy debates? *PLAAS Institute for Poverty, Land and Agrarian Studies, Working Paper 44*. University of the Western Cape. Available at: www.plaas.org.za/plaaspublication/wp44_informalsector_skinnerhaysom [Accessed 24 Jan. 2018].

Smith, J. and Abrahams, M. (2016). *2016 PACSA food price barometer annual report*. Pietermaritzburg. Available at: www.pacsa.org.za/images/food_barometer/2016/2016_PACSA_Food_Price_Barometer_REDUCED.pdf [Accessed 22 Jan. 2018].

Swindale, A. and Bilinsky, P. (2006). *Household dietary diversity score (HDDS) for measurement of household food access: Indicator guide (v.2)*. Washington, DC: FHI 360/FANTA.

UNICEF, (2008). *State of the world's children, 2009*. New York, NY: UNICEF.

18 Food poverty in Epworth, Zimbabwe

Godfrey Tawodzera and Easther Chigumira

Introduction

For close to two decades now food poverty and insecurity in Zimbabwe has consistently been high. The recent Global Hunger Index found current levels of undernourishment of 44.7%, with very little change since the early 1990s (IFPRI/Welthungerhilfe/ConcernWorldwide 2017). Levels of food insecurity in urban areas are high. Tawodzera *et al.* (2016) found that only 17% of households in the low-income areas of Harare could be classified as food secure. While the later statistic paints an alarming picture of poverty and food insecurity in urban Zimbabwe, particularly the large urban centres, little is still known about the conditions that exist in the medium-sized and peri-urban centres like Epworth, where most of the future urban growth is predicted to occur. The objective of this chapter is therefore to explore the nature and extent of food poverty in Epworth and to examine factors that are critical to household food security in the area. To achieve this objective, the chapter presents findings from two surveys carried out in Epworth in 2016. Located 20 kilometres to the east of Harare city centre, Epworth is home to approximately 167 462 people (Zimbabwe National Statistics Agency 2012), more than two-thirds of whom live in informal conditions where poverty is endemic (Zimbabwe National Statistics Agency 2013).

This chapter contextualizes the food insecurity of Epworth's residents within the wider political and economic context of Zimbabwe, arguing that the food insecurity described in the chapter is shaped by national factors driving the food system and economy as much as household-scale factors. Following a description of the characteristics of food insecurity in Epworth, the chapter critically engages with the relationship between poverty and food insecurity. The chapter argues that household food insecurity is the outcome of multidimensional poverty experienced at both the household and the neighbourhood scales.

The country context

The Introduction of this book provides a description of Epworth itself, but for this chapter it is important to provide a contextual understanding of factors shaping the national food environment. One of the important factors in this regard has been the controversial Fast Track Land Reform Programme which was implemented countrywide from the year 2000. The programme disrupted farming operations, affecting overall food production in the country (Sachikonye 2005). The new land beneficiaries, generally lacking farming experience and sufficient financial support, have struggled to produce enough food to feed the country (Muziri 2009). Successive droughts and policy inconsistencies by the government have worsened the economic and food security situation in the country. Zimbabwe, previously the breadbasket of the Southern African Development Community member-states, has thus become a perennial net importer of food. While the country has been facing all these challenges, poverty levels have been increasing, as the 2003 Poverty Assessment Study Survey (PASS) indicated that 63% of households in Zimbabwe were living below the poverty consumption line (Government of Zimbabwe 2006). By 2008, it was estimated that close to 80% of the population was living on less than US$2 per day (Chimhowu 2009). The situation has not improved much. In 2015, the government estimated that close to 70% of the country's population was living below the total consumption poverty line (Government of Zimbabwe 2016). Of late, a number of policies introduced by the government have further negatively impacted on food availability in the country – for example the Statutory Instrument 64 of 2016 banned the importation of certain food products, limited the importation of others, and imposed heavy duties on others (USAID 2016). This resulted in price increases of certain foodstuffs and temporary shortages of others, further worsening food insecurity in a country reeling from high unemployment and low wages. Food poverty in Epworth should thus be understood in the context of the aforementioned dynamics operating at the larger national scale.

Research method

The Consuming Urban Poverty (CUP) study in Epworth employed a mixed-methods approach involving household semi-structured qualitative interviews and a standardized quantitative household food security questionnaire survey. Using statistical formule, and working with a total population of 47 028 households (Zimbabwe National Statistics Agency 2012), and a 5% standard error at a 95% level of probability, the appropriate sample size was determined to be 381 households. However, given that the census figure for the total number of households in Epworth was almost four years out of date, the sample size was increased to 481 households. In the end, 483 household questionnaires were administered.

The sample was drawn from only three of the seven wards in Epworth. This decision was made in part due to safety concerns of working in the remaining four wards at the time. The three wards selected were a relatively poor ward (Ward 6), a better-off ward (Ward 4), and one lying between these (Ward 2). The household survey therefore represents only the numbers in these three wards and not the whole of Epworth. However, the differences in incidence of poverty between Epworth's seven wards was just four percentage points, with the poorest Ward (Ward 5) having a poverty prevalence of 66.6% and the least poor (Ward 4) having a poverty prevalence of 62.5%. The poverty prevalence of Epworth as a whole was 64.5%. The population of Epworth is therefore relatively homogenous (Zimbabwe National Statistics Agency 2012: 212). The sample sizes in each ward were proportional to the number of households in each ward (see Table 18.1). A further limitation of the data is that the households in the three wards were not selected with an equal probability, so the percentage estimates for all the three wards (together) are not perfectly representative of the numbers in the total population in these three wards.

To secure a representative sample of the three wards, a systematic sampling approach with a random start was adopted, following the principles of probability sampling (Czaja and Blair 1996). Informed by the geography of the ward, the enumeration team decided on which entry points were the most appropriate and designated key entry streets were identified. Enumerators entered the area at multiple points. They used a random number generator app on their tablets (a range from one to four) to select the house with which they would begin the systematic sample. Once the sampling intervals were computed from the total number of households in each of the selected wards, every 38th (Ward 2), 44th (Ward 4), and 31st (Ward 6) house along the road was sampled. Within households, an adult household member above the age of 18 years who was either the household head or could speak for the household was engaged for the interview.

The core focus of the questionnaire was the FANTA food security indicators and the Lived Poverty Index. The indicators include the Household Food Insecurity Access Prevalence Indicator (HFIAP), the Household Dietary Diversity Scale (HDDS), and the Months of Adequate Household Food Provisioning Indicator (MAHFP). Further details on the FANTA indicators can be found in Swindale and Bilinsky (2006), Coates *et al.* (2007), and Bilinsky and Swindale (2010).

Fifteen semi-structured interviews were carried out. Households were chosen by the snowball method and new households were chosen to interview until they appeared not to be yielding new information. This is the concept of 'data saturation' developed by Glaser and Strauss (1967). Information from these interviews was used to inform aspects of the quantitative survey.

Table 18.1 Total number of questionnaires administered in Epworth, by ward

Survey area	*No. of households*	*Sample no. of households*	*%*
Ward 2	4280	113	23.4
Ward 4	6159	141	29.2
Ward 6	7151	229	47.4
Total	**17590**	**483**	**100.0**

Study results

Sample characteristics

More than half the sample population (51%) were females while 49% were males. This echoed the national sex ratio of 97 males for every 100 females. However, 79% of sampled households were male headed, far higher than the national proportion of 65% (Zimbabwe National Statistics Agency 2012). About half (51%) of the sample population was below 20 years. The average age of the sample was 22 years while that for household heads was 42 years. The youngest household head was aged 20 years and the oldest was 92 years. The sample population was well educated, with just 4% of household heads having had no formal schooling, while 39% had completed at least a high school education, and a further 40% had some high school education.

Close to half of the sample population were students (47%). Of the non-student population, 29% identified as self-employed, 13% were working full time, 13% working part time, 26% were unemployed, and 17% identified as housewives. The low proportion of respondents in full-time employment is indicative of the current Zimbabwean economy where the formal sector is limited, a context in which self-employment becomes increasingly important (Barba-Sanhez and Atienza-Sahuquilo 2017). Given the methodological diversity in approaches to measuring unemployment in Zimbabwe (Chumia 2014), it is difficult to comment on whether unemployment is above average. However, the poverty data presented in this chapter indicate that the majority of households eke out marginal existences in Epworth.

A characteristic of most distressed economies is that people migrate to work elsewhere as jobs become scarce in their environment (Feser and Sweeney 1999). In the Epworth survey, a total of 2253 household members were identified in 483 households. Of these individuals, just 2% were identified as household members working elsewhere. Just 22 households (5%) listed any income from remittances of cash. In a country where a high percentage of the workforce has migrated to neighbouring countries, the low proportion of household members living away from home in Epworth is a paradox. While poverty is generally seen as a trigger for out-migration, Guriev and Vakulenko (2015) argue that acute poverty may actually be debilitating, locking the poor in a geographical poverty trap from which they are unable to escape. It is therefore possible that the acute poverty levels among households in Epworth are precluding members from migrating as they cannot raise the resources needed to move. Most Epworth households are thus being deprived of income that migrant workers could potentially generate to ease their food problems.

Household poverty

The CUP survey collected data on household income and expenditure. Almost half of the households (46%) reported monthly incomes of US$100 or less, and

a further 26% reported incomes of between US$101 and US$200. Only 5% of households reported incomes of over US$500. The total household mean monthly income was US$185.55. The corresponding mean household expenditure of sampled households was US$143.80, with a mean household monthly food expenditure of just US$53.82. In July 2016 the Consumer Council of Zimbabwe stated that the monthly food basket for a family of six in Zimbabwe was US$567 (Mlilo 2016). The extreme gap between stated expenditure on food and the national food basket cost suggests significant food and nutrition insecurity.

National governments often depend on income and expenditure figures as proxy indicators for food insecurity. However, income is only one component of poverty. To fully understand poverty levels in Epworth, this study used the Lived Poverty Index (LPI), a multidimensional measure that captures current poverty dynamics as indicated by the failure of the household to access a variety of goods and services, namely access to food, clean water, necessary medical care, fuel for cooking, and cash. The LPI is computed into a composite score ranging from 0.00 to 4.00, with a score of 4.00 indicating that households were always without all the categories. The average LPI in Epworth was 1.39. This is slightly above the 2014/2015 national average of 1.36 (Mattes *et al.* 2016).

More than half of the households (53%) indicated that they had gone without enough food several times, many times, or always. Almost half (46%) had gone without clean water, 30% without medicine or medical care, 46% without enough fuel to cook, and 61% without cash with the same frequency. These multiple deprivations are indicative of an environment where local authorities have failed to maintain a stable economic environment for households to thrive. The situation in Epworth is similar to what Sen (1999) regards as an entitlement failure where people are unable to secure their means of subsistence because their entitlements (e.g. a living wage, insurance, health) are no longer legally guaranteed. The concatenating impact of lack of access to clean water and fuel to cook with means that food access limitations are compounded with utilization limitations (see Chapter 3).

Levels of household food security

Food security is complex, multi-faceted, and difficult to measure. This is because it does not have a single compound measure (Ballard *et al.* 2013). Survey results indicate high levels of food insecurity among households in Epworth. The HFIAP measure found that only 8% of surveyed households were food secure. The remaining households were mildly food insecure (5%), moderately food insecure (21%), and severely food insecure (67%). In 2008, at the height of the country's economic crisis, a study in Harare which had a pro-poor sampling strategy, including households from Epworth, found just 2% of sampled households were food secure, 3% were mildly food insecure, 24% moderately food insecure, and 72% severely food insecure (Tawodzera *et al.* 2012). A similar survey was conducted in 2012 during the tenure of the Government of

National Unity and found 10% of sampled households food secure, 7% mildly food insecure, 20% moderately food insecure, and 63% severely food insecure (Tawodzera *et al.* 2016).

The apparent decline in food insecurity between 2012 and 2017 may be reflective of the country's worsening economic situation. The survey found that some households in Epworth were even consuming just a single meal per day. One respondent explained the reality of this situation:

> Right now my husband is not going to work, so in the morning we might not have anything to eat and strive to get to midday on empty stomachs. We then roast peas and drink water afterwards. We then wait for sadza in the evening on those days that mealie meal is available. If we have no mealie meal, we eat roasted peas again for supper then go to sleep.
>
> Interview 1, Ward 1, Epworth, 25 June 2016

The struggle to bring food on the table was being exacerbated by the general shortage of cash in the country. Throughout 2016, the Reserve Bank of Zimbabwe periodically reviewed cash withdrawal limits downwards (e.g. Mangundya 2016). While the use of the US$ may have stabilized the economy, increased food imports, and made food more available on the market, it has also resulted in a shortage of cash as the US$ is not minted in the country and there is no national currency in circulation. Added to that problem is the fact that the economy is not performing well enough to produce exports that generate foreign currency (Reserve Bank of Zimbabwe 2016). A respondent in Ward 2 explained this more succinctly:

> I am sure that you are aware that these days, some could be having a small income but when they go to the bank they can't access it because the cash often runs out. Because almost every day, you have to travel to and from the bank using money for transport, it causes pressure and shortages here at home since you then cannot buy enough food. In some instances, there are debts of money borrowed from neighbours so as to buy food. These debts need to be repaid such that all the obtained cash is depleted and the borrowing cycle is repeated.
>
> Interview 8, Ward 2, Epworth, 26 June 2016

The problem of cash shortages is magnified in Epworth as there is no bank in the area and anyone requiring banking services has to commute to Harare. In an environment where bank withdrawal limits were as low as US$50 and where multiple journeys would have to be made before a successful withdrawal, commuting to access banking services in Harare was rather wasteful of the meagre cash resources that could have gone to purchase food.

The severity of food insecurity among households in Epworth was also reflected in narrower diets. No household in the survey reported consuming from all 12 food groups. The majority of the households (80%) reported consuming foodstuffs from five food groups or less. Only 20% of the households were consuming from six food groups or more. A HDDS score of less than six out of 12 is generally considered to be a proxy for malnutrition. The mean HDDS score was 4.12 out of 12.00. Given these statistics, one can justifiably conclude that households in Epworth had nutritionally inadequate diets which are detrimental to their health. As Powell *et al.* (2017) point out, under-nutrition and micro-nutrient deficiency remain one of the challenges to global development, particularly in poor areas, and Epworth is no exception. Respondents to the in-depth interviews testify to the bankruptcy of their diets:

> We normally eat foods with lots of starch and fats. It is difficult to include fruit in our diet because we don't have the money. Here and there you will find money for meat but it's usually on rare occasions. Once we have exhausted the monthly allocation for meat, we then substitute it with matemba and vegetables.
>
> Interview 10, Ward 5, Epworth, 26 June 2016

These results confirm observations made by ZimVac (2009) in their Urban Food Security Assessment that poor households in urban areas were consuming less diversified diets. In that study, close to half of the urban households surveyed (46.52%) were consuming what could be classified as poor or borderline diets. Unfortunately, not many follow-up studies have been done, especially in small urban centres, to further interrogate the depth of this dietary inadequacy.

What were the types of foods that were being consumed?

All but one of the 483 sampled households indicated that they had consumed vegetables within the last 24 hours. These were the most commonly consumed foods, followed by cereal products (95%), sugar or honey (40%), foods made with fats/oil (32%), and meat or chicken (28%). The least consumed products were roots or tubers (11%), fruits (17%), and dairy products (15%). Ironically, yet unsurprisingly, these least-consumed foods are generally known to promote health. It has been shown in studies that the consumption of fruits (and vegetables) can prevent oesophageal, stomach, pancreatic, bladder, and cervical cancers (Oguntibeju *et al.* 2013). Dairy products have been associated with lower incidence of certain types of diabetes and positive effects on blood pressure (Visioli and Strata 2014). The fact that Epworth residents were consuming less of these healthy foods is worrying, given that this population is the least able to bear the cost of the potential health implications of this diet.

In addition to low dietary diversity, the variability of food supplies within households in Epworth is also of concern. More than three-quarters (78%) of surveyed households reported experiencing months of inadequate food provisioning in the past 12 months. A mean score of 8.89, representing almost nine months of inadequate food provisioning, was recorded. The months in which most households were inadequately provisioned were from August to December. This trend can be explained by the fact that though Epworth is urban, it is also heavily interconnected to rural food cycles. It therefore experiences notable increases in food prices during those months where rural food production is low. From late December, the situation begins to stabilize as food grown in the rural areas and in the peri-urban areas begins to ripen. This increases food that is available on the urban market for purchase by households that have enough resources. In some cases, the increased inflow of food to the urban market during this period tends to depress food prices, hence improving urban households' economic access.

The food problems in Epworth are not just transient. They are a reflection of deep-seated problems ranging from lack of proper housing, inadequate water, poor sanitation facilities, and poor roads. In broader terms, Epworth residents are suffering because of 'capability deprivation.' As Sen (1999) points out, people generally suffer because the political, social, and economic structures that safeguard and enhance their freedom and agency have been compromised. The collapse of these structures thus means that people are likely to experience problems and go hungry. It is no wonder therefore that close to half (42%) of the respondents in Epworth indicated that their household economic conditions were now much worse compared to a year ago, while 30% regarded their condition as just worse. Only 13% considered their situation to be much better than a year ago.

Where do Epworth residents buy their food?

According to survey results, the greater proportion of households buy from vendors/traders/hawkers in Epworth (74%), followed by markets in Epworth (63%), vendor/trader/hawkers in Harare (56%), and from tuck-shops in Epworth (52%). Households indicated buying from these food sources at least five days per week. Thus the food sources in the area are dominated by informal food sources. The other dominant food sources are butcheries (Epworth) where 53% of the households indicated buying once a week. Supermarkets, on the other hand, are patronized mostly once a month by 63% of the households, while wholesalers in Harare (72%) and wholesalers in Epworth (60%) are also patronized once a month. The trend that emerges here is that informal food sources are patronized more frequently and closer to home than the formal food sources that are patronized less frequently and are generally further from Epworth.

Linking food insecurity and poverty in Epworth

Food poverty is entrenched in Epworth as indicated in the preceding discussion. This section will briefly dwell on the key variables influencing this food poverty. First, income is crucial. Several studies have shown that access to food and expenditure on food in poor areas depends on whether households in urban areas have enough income to purchase food (Hoyos and Mevedev 2009). Results in this survey show that even within a predominantly poor sample, households with more income were generally more food secure, with the mean income of food-secure households being US$377.65, and that of severely food-insecure households being US$148.55. The low-income levels among most of Epworth's households therefore mean that they are more vulnerable to food insecurity. Any meaningful attempt to deal with food poverty in the area thus has to make the improvement of income central.

However, it is clear that the determinants of food insecurity extend beyond income poverty alone. As Table 18.2 indicates, there is a strong correlation between food security outcomes and LPI. No household with an LPI over 2.00 was food secure or just mildly food insecure. Food insecurity is the outcome of multiple deprivations that shape food accessibility and utilization, thus entrenching food insecurity in households lacking sufficient money, water, fuel, and healthcare. These limits on accessibility and utilization lead to the low dietary diversity found in Epworth. As long as households in Epworth continue to struggle accessing adequate and varied diets, they are most likely to be food insecure.

Conclusion

Epworth's residents experience acute food poverty. The majority of the people in the area live in informal settlements, with little or no access to water, electricity, sanitation, or serviced roads. As Table 18.2 indicates, these infrastructural deficiencies shape the ability of households to meet their food security needs in ways that go beyond simple income poverty. This therefore implies that if food security as a manifestation of poverty is to be addressed, it will require more systemic approaches than strategies to increase household income and improve livelihoods alone.

The CUP household survey identified the informal sector and self-employment to be important sources of employment and income within Epworth. It also found informal food retailers to be an important source of food for Epworth's residents. However, this double dependency on informality has food security implications. The limited and inconsistent incomes generated

Table 18.2 Number of households in each food security category by LPI score

	LPI score			
	0.00–1.00	*1.01–2.00*	*2.01–3.00*	*3.01–4.00*
Food secure	31	6	0	0
Mildly food insecure	19	3	0	0
Moderately food insecure	45	47	6	0
Severely food insecure	79	158	81	2

through informal sector employment and self-employment impact households' abilities to purchase sufficient nutritious food. This, combined with the multiple deprivations identified by the LPI tool, limits food security and potential dietary diversity. Because of the marginal incomes earned by informal sector food retailers in areas serving predominantly poor populations, retailers are effectively bound by selling what they know their consumers will purchase. This, combined with limited access to water, sanitation, and fuel, means that they are more likely to sell a limited range of foods. This vicious cycle is reinforcing limited dietary diversity, which raises concerns about not only the current food poverty issues, but also future implications for diseases associated with narrow diets as well as the future disease costs for a population that is poor and unable to deal with the consequences. Dollarization, while stabilizing the economy, is also having a negative effect on household food security as households struggle to access cash.

In order to address food poverty in Epworth it is essential to have a multi-pronged strategy that aims at alleviating household income poverty through improving livelihoods, at the same time as improving access to basic infrastructure for both households and economic enterprises. This will ensure that food is not just economically and physically accessible, but is also able to be utilized by households to improve food security and long-term health.

References

Ballard, T.J., Kepple, A.W. and Cafiero, C. (2013). *The food insecurity experience scale: Development of a global standard for monitoring hunger worldwide, technical paper.* Rome: FAO. Available at: www.fao.org/economic/ess/ess-fs/voices/en/ [Accessed 20 June 2017].

Barba-Sanhez, V. and Atienza-Sahuquilo, C. (2017). Entrepreneurial motivation and self-employment: Evidence from expectant theory. *International Entrepreneurial Management Journal*, 13(4), pp. 1097–1115.

Bilinsky, P. and Swindale, A. (2010). *Months of adequate household food provisioning (MAHFP) for measurement of household food access: Indicator Guide (v. 4).* Washington, DC: FHI 360/FANTA.

Chimhowu, A. (2009). *Moving forward in Zimbabwe: Reducing poverty and promoting productivity.* Manchester: Brooks Poverty Institute.

Chumia, S. (2014). Is Zimbabwe's unemployment rate 4%, 60% or 95%? Why the data is unreliable. *Africa Check*, 1 October 2014. Available at: https://africacheck.org/reports/is-zimbabwes-unemployment-rate-4-60-or-95-why-the-data-is-unreliable/ [Accessed 25 Jan. 2018].

Coates, J., Swindale, A. and Bilinsky, P. (2007). *Household food insecurity access scale (HFIAS) for measurement of food access: Indicator guide (v.3).* Washington, DC: FHI 360/FANTA.

Czaja, R. and Blair, J. (1996). *Designing Surveys: A guide to decisions and procedures.* Thousand Oaks, CA: Pine Forge Press.

Feser, E.J. and Sweeney, S.H. (1999). *Out-migration, population decline, and regional economic distress.* Washington, DC: Department of City and Regional Planning, Economic Development Administration, United States Department of Commerce.

Glaser, B.G. and Strauss, A. (1967). *The discovery of grounded theory: Strategies for qualitative research.* Chicago, IL: Aldine.

Government of Zimbabwe, (2006). *Poverty Assessment Study Survey, Main Report; Ministry of Public Service, Labour & Social Welfare.* Harare: Government Printers.

Government of Zimbabwe, (2016). *Zimbabwe interim poverty reduction strategy paper (I-PRSP): 2016–2018.* Ministry of Finance & Economic Development. Available at: www.veritaszim.net/sites/veritas_d/files/Interim%20Poverty%20Reduction%20Strategy%20Paper%20(I-PRSP)%20 2016%20-%202018.pdf [Accessed 17 May 2017].

Gurieve, S. and Vakulenko, E. (2015). Breaking out of poverty traps: Internal migration and interregional convergence in Russia. *Journal of Comparative Economics*, 43, pp. 633–649.

Hoyos, R.E. and Mevedev, D. (2009). Poverty of higher food prices, a global perspective. *Policy Research, Working Paper 4887.* Washington, DC: The World Bank.

IFPRI/Welthungerhilfe/ConcernWorldwide, (2017). *2016 global hunger index: The inequalities of hunger.* Washington, DC/Dublin/Bonn: IFPRI.

Mangundya, J.P. (2016). *Measures to deal with cash shortages whilst simultaneously stabilising and stimulating the economy.* [press statement]. Reserve Bank of Zimbabwe.

Mattes, B.R., Dulani, B. and Gyimah-Boadi, E. (2016). *Africa's growth dividend? Lived poverty drops across much of the continent.* Afrobarometer Policy Paper, No. 29, January 2016. Available at: http://afrobarometer.org./sites/default/files/publications/Policy%20papers/ab_r6_policypaperno29_lived_poverty_declines_in_africa_eng.pdf [Accessed 1 May 2018].

Mlilo, B. (2016). Food basket hits $567. *Chronicle.* Available at: www.chronicle.co.zw/food-basket-hits-567-price-hikes-put-strain-on-average-consumer/ [Accessed 25 Jan. 2018].

Muziri, T. (2009). Farm management challenges for a country in transition: The case of Zimbabwe. *Journal of International Farm Management*, 4(4), pp. 1–14.

Oguntibeju, O.O., Truter, E.J. and Esterhuyse, A.J. (2013). The role of fruit and vegetable consumption in human health and disease prevention, diabetes mellitus-insights and perspectives. doi:10.5772/50109

Powell, B., Kerr, R.B., Young, S.L. and Johns, T. (2017). The determinants of dietary diversity and nutrition: Ethnonutrition knowledge of local people in East Usambara Mountains, Tanzania. *Journal of Ethnobiology and Ethnomedicine*, 13(23), pp. 1–12.

Reserve Bank of Zimbabwe, (2016). *Public notice on export bonus scheme which is supported by bond notes: Strategy to stimulate export production to maintain and sustain the multi-currency system in Zimbabwe*, Harare: Reserve Bank of Zimbabwe.

Sachikonye, L. (2005). The land is the economy: Revisiting the land question. *African Security Review*, 14(3), pp. 31–44.

Sen, A. (1999). Development as freedom. New York, NY: Oxford University Press.

Swindale, A. and Bilinsky, P. (2006). *Household dietary diversity score (HDDS) for measurement of household food access: Indicator guide (v.2).* Washington, DC: FHI 306/FANTA.

Tawodzera, G., Riley, L. and Crush, J. (2016). The return of food: Poverty and urban food security in Zimbabwe after the crisis. *AFSUN Food Security Series*, (22), Cape Town and Waterloo. Available at: www.afsun.org/wp-content/uploads/2016/06/AFSUN22.pdf [Accessed 20 June 2017].

Tawodzera, G., Zanamwe, L. and Crush, J. (2012). *The state of food insecurity in Harare, Zimbabwe.* AFSUN Food Security Series, (13), Cape Town. Available at: www.afsun.org/wp-content/uploads/2016/06/afsun13.pdf [Accessed 20 June 2017].

USAID, (2016). *Draft Report on the Review of the Impact of Statutory Instrument 64 of 2016.* Available at: http://pdf.usaid.gov/pdf_docs/PA00MDKJ.pdf [Accessed 17 May 2017].

Visioli, F. and Strata, A. (2014). Milk, dairy products and their functional effects in humans: A narrative review of recent evidence. *American Society for Nutrition*, 5(2), pp. 131–143.

Zimbabwe National Statistics Agency, (2013). *Poverty and Poverty Datum Line Analysis in Zimbabwe 2011/12, ZimStat.* Available at: www.zimstat.co.zw/sites/default/files/img/publications/Finance/Poverty2011.pdf [Accessed 20 May 2017].

Zimbabwe National Statistics Agency, (2012). *Zimbabwe population census 2012: National Report, ZimStat, Harare.* Available at: www.zimstat.co.zw/sites/default/files/img/National_Report.pdf [Accessed 20 May 2017].

Zimbabwe Vulnerability Assessment Committee, (ZimVac),.(2009). *Urban Food Security Assessment*. SIRDC. Available at: http://documents.wfp.org/stellent/groups/public/documents/ena/wfp197654.pdf?iframe [Accessed 17 May 2017].

Index

Note: Page numbers in *italic* indicate a figure and page numbers in **bold** indicate a table on the corresponding page.

For Product Safety Concerns and Information please contact our EU representative GPSR@taylorandfrancis.com
Taylor & Francis Verlag GmbH, Kaufingerstraße 24, 80331 München, Germany

www.ingramcontent.com/pod-product-compliance
Ingram Content Group UK Ltd.
Pitfield, Milton Keynes, MK11 3LW, UK
UKHW020014200926
475702UK00014B/1886
* 9 7 8 0 3 6 7 5 8 7 5 6 7 *